Introduction to Construction Contract Management

This book is an introduction to construction contract administration and management, covering the delivery and execution stage of a construction project and the various issues which the contract administrator needs to proactively manage. It can therefore be used as a contract administrator's resource book covering what needs to be done (and why) to keep a construction project on track from a commercial and contractual perspective. It is particularly appropriate for students and new practitioners from varied construction professions and whilst it covers domestic (UK) projects, it will be particularly useful for those studying and working on international projects where terminology, procedures and legal systems may differ from the UK.

The content is split into four parts and is subdivided into easy-to-read chapters replicating the timeline of a project during the construction stage:

- Part A covers initiating the construction stage, project delivery mechanisms, contract administration and health and safety management;
- Part B covers managing the construction stage, contractor performance and relationship management;
- Part C covers finalising the construction stage, project completion and close-out;
- Part D covers claims and disputes.

Introduction to Construction Contract Management will be particularly useful for students enrolled on global construction programmes together with international distance learning students and non-cognate graduates starting out on an international career in construction contract administration and quantity surveying.

Brian Greenhalgh FRICS FCIOB has worked for many years as a commercial / contract manager for major client organisations in the MENA region. He was also formerly a principal lecturer in quantity surveying and construction project management at Liverpool John Moores University, with responsibility for postgraduate programmes in quantity surveying and construction project management.

Introduction to Construction Contract Management

Brian Greenhalgh

LONDON AND NEW YORK

First published 2017
by Routledge
2 Park Square, Milton Park, Abingdon, Oxon OX14 4RN

and by Routledge
711 Third Avenue, New York, NY 10017

Routledge is an imprint of the Taylor & Francis Group, an informa business

British Library Cataloguing-in-Publication Data
A catalogue record for this book is available from the British Library

Library of Congress Cataloging in Publication Data
Names: Greenhalgh, Brian, author.
Title: Introduction to construction contract management / Brian Greenhalgh.
Description: Abingdon, Oxon ; New York, NY : Routledge is an imprint of the Taylor & Francis Group, an Informa Business, [2016] | Includes bibliographical references and index.
Identifiers: LCCN 2016007639| ISBN 9781138844148 (hardback : alk. paper) | ISBN 9781138844179 (pbk. : alk. paper) | ISBN 9781315730585 (ebook : alk. paper)
Subjects: LCSH: Building—Superintendence. | Construction contracts. | Project management.
Classification: LCC TH438 .G6735 2016 | DDC 692/.8—dc23
LC record available at https://lccn.loc.gov/2016007639

ISBN: 978-1-138-84414-8 (hbk)
ISBN: 978-1-138-84417-9 (pbk)
ISBN: 978-1-315-73058-5 (ebk)

Typeset in Sabon
by Swales & Willis Ltd, Exeter, Devon, UK

To Sheila, Neil and Kara, Simon and Steph –
and of course Morgan and Bobby

Contents

Figures

Tables

Preface

In January 2011 Routledge published the sister book to this, *Introduction to Building Procurement*, which I co-authored with Dr Graham Squires of the University of the West of England. In January 2013 Routledge also published *Introduction to Estimating for Construction*, and once the dust had settled on these it was clear to me that there was a circle to complete. The missing arc was in the construction stage contract management, since the first two books covered the pre-contract area of construction commercial management. This book now hopefully completes that arc and the three books together give the student an easy to read introduction to the whole area of commercial management of construction. A 'trilogy' if you will!

This is not a textbook on construction law. I have deliberately tried to avoid any references to legal cases, statutes, contract clause analysis etc. and also tried not to use legal language unless absolutely necessary to illustrate points or give the general principle – keep it simple is the mantra. The reader is encouraged to understand the big picture and then drill down to the level of detail required from other, more specialised texts. Therefore, this *is* a discussion of the issues relating to construction contract administration and management which covers both domestic and international projects. It can therefore be used as a contract administrator's resource book covering what needs to be done (and why) to keep a construction project on track from a commercial and contractual viewpoint. It is also particularly appropriate for students – and for new practitioners from the other construction professions – to find out what this commercial management business is all about.

In keeping this textbook down to a manageable size, I hope I have achieved it without impairing the quality or glossing over important issues.

As with the previous two textbooks, the contents and structure of this book are slightly different from traditional texts on the subject matter and again I make no apology for this. The main reason is that the construction industry, both domestically and internationally, has changed quite considerably over the last twenty years – procurement systems with more integrated contractor involvement are in more common use, therefore the traditional 'master and servant' relationship is reducing – although try telling that to some clients.

Brian Greenhalgh
January 2016

Acknowledgements

I would like to express my great thanks to my colleagues and friends, Paul Winfield, Kelvin Hughes and Mike Testro, who have helped in the writing of this book by providing written material and real life examples or by reading over the material and offering practical advice where they thought it was clearly required.

Abbreviations

ABC	Anti-bribery and corruption
ADR	Alternative dispute resolution
ANB	Adjudicator nominating body
BATNA	Best alternative to a negotiated agreement
BIM	Building Information Modelling
BLS	Baseline schedule
BOQ	Bill of Quantities
BS	British Standard
CAD	Computer aided design
CAR	Contractor's All Risks Insurance
CDB	Combined dispute board
CDM	Construction (Design and Management) Regulations
CEBE	Centre of Excellence for the Built Environment
CEO	Chief executive officer
CEP	Contract execution plan
CESMM	Civil Engineering Standard Method of Measurement
CIOB	Chartered Institute of Building
CPARS	Contractor Performance Assessment Reporting System
CPM	Critical path method
CPR	Civil Procedure Rules
CSF	Critical success factor
DAB	Dispute Adjudication Board or Dispute Avoidance Board
DB	Design–Build
DLP	Defects liability period
DMS	Document Management System
DRB	Dispute Review Board or Dispute Resolution Board
ECC	Engineering and Construction Contract
ECI	Early contractor involvement
EN	European norm
EOT	Extension of time
EPC	Engineer, procure, construct
FAST	Function Analysis and System Technique
FEED	Front end engineering design
FIDIC	International Federation of Civil Engineers
FM	Facilities management
GCC	Gulf Cooperation Council

HEMP	Hazard Effect Management Programme
HSE	Health, Safety and Environment
HVAC	Heating, ventilating, air conditioning
ICC	Infrastructure Conditions of Contract
ICE	Institution of Civil Engineers
ICT	Information and communications technology
ICV	In-country value
JCT	Joint Contracts Tribunal
KPI	Key performance indicator
LAD	Liquidated and ascertained damages
LD	Liquidated damages
LOA	Letter of Award
LTI	Lost time incident
MAS	Materials approval system
MEP	Mechanical, electrical and plumbing
NEC	New Engineering Contract
NRM	New Rules of Measurement
NTP	Notice to Proceed
OGC	Office of Government Commerce
O&M	Operation and maintenance
PC	Prime cost
PCSA	Pre-construction services agreement
PEP	Project execution plan
PI	Professional indemnity insurance
PM	Project manager
PMP	Project management plan
PMS	Project management system
PO	Purchase Order
PPE	Personal protective equipment
PSA	Professional services agreement
PTE	Pre-Tender Estimate
QA	Quality assurance
RFI	Request for information
RFP	Request for proposal
RFQ	Request for quotation
RIBA	Royal Institute of British Architects
RICS	Royal Institution of Chartered Surveyors
SBC	Standard Building Contract
SCL	Society of Construction Law
SIA	Schedule impact assessment
SLA	Service Level agreement
SMM	Standard method of measurement
SOR	Schedule of record
SOW	Scope of work
TCC	Technology and Construction Court
TIA	Time impact analysis
TOR	Terms of reference
UNCITRAL	United Nations Commission on International Trade Law

VA	Value analysis
VAT	Value added tax
VE	Value engineering
VECP	Value engineering change proposal
VO	Variation Order
VTC	Variation to Contract
WBS	Work breakdown structure

Introduction

As stated in the Preface, this book is an *Introduction* to construction contract administration and management, covering the delivery and execution stage of a construction project and the various issues which the contract administrator needs to proactively manage. Figure 0.1 gives a comprehensive list of the various project management issues throughout each stage of a construction project, with the issues covered in this book in the shaded area.

However, as stated above, this is a book on contract administration rather than project management, so the discussions in the chapters will concern how these issues affect the contract rather than how they are dealt with by the project manager. Clearly, there will be some overlap (for example programming and scheduling), so some explanation of the project management techniques is necessary in these cases.

So what then, is construction contract administration? Figure 0.2 gives an illustration of the four decisions that must be made on a project. When the design decisions have been made, normally tempered by what the employer can afford, this creates the 'scope of works'. Time and schedule decisions are then made by both the employer (required completion date) and the contractor (using critical path techniques) and together they will make the project configuration. Procurement decisions are really the 'how'? – do we employ the contractor early? What supervision consultants do we need? etc. All together, the project scope of works and procurement decisions will give the contract configuration, and it is the management of this package that this book is about – managing and administering the package to ensure the project achieves its objectives.

There are many different forms of contract in common use in the global construction industry and the terminology and roles of the participants can be slightly different in each of them. This has been my challenge in this book – to construct a clear account of the various issues and procedures in construction contract administration which can be applicable across many different forms of contract and legal jurisdictions without making it unnecessarily complex or unreadable by repeating the different terms for what can be essentially the same role.

The obvious example is that of the leader of the project. The major standard forms of contract use the terms Engineer, Contract Administrator, Project Manager etc. and some internal client-based contracts use the term Contract Holder, so it is not surprising that confusion is created.

Therefore, for the purposes of simplicity, I have generally used the term PM / engineer to describe what this person would do in relation to the issue under discussion. All construction projects go through the same stages and series of issues

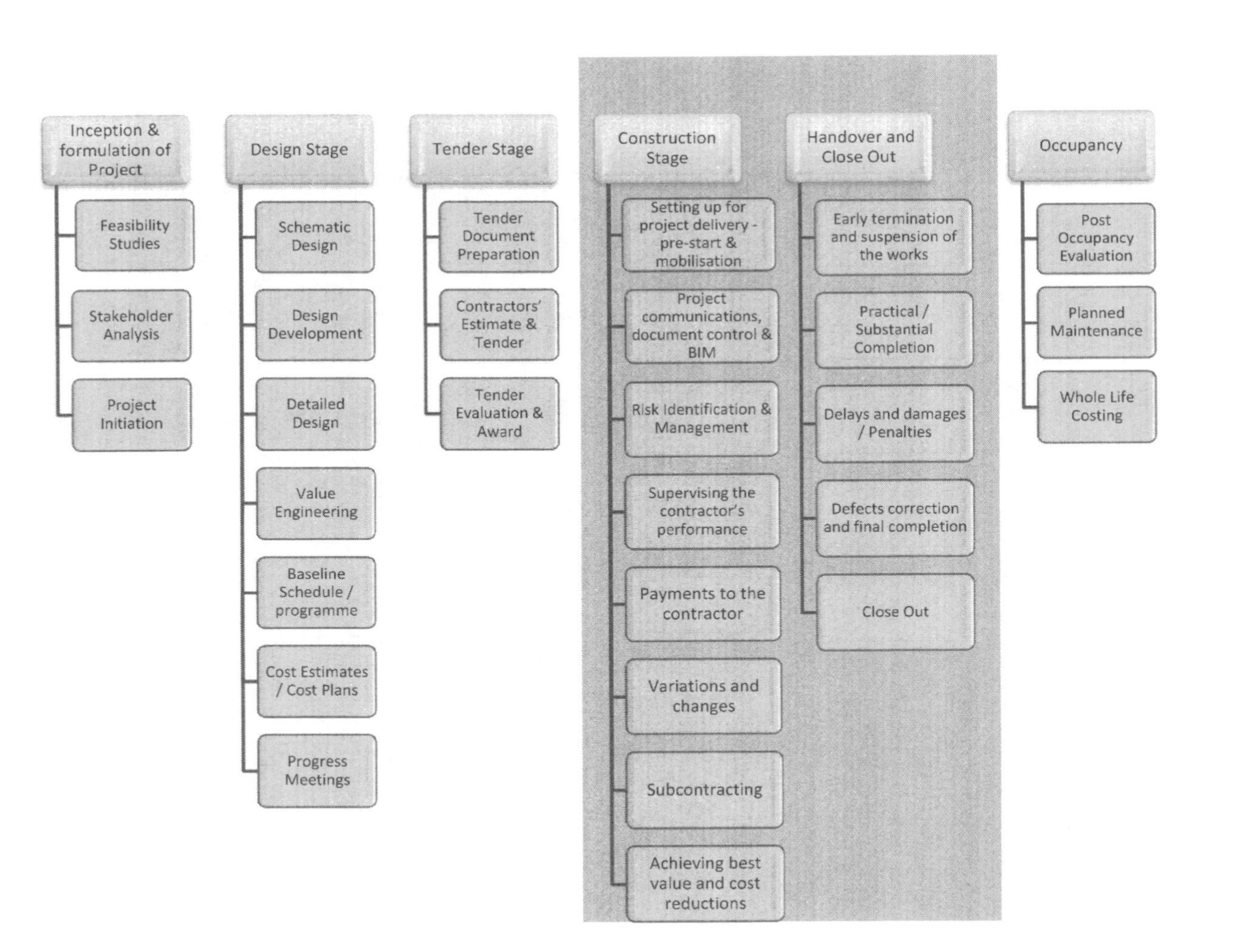

Figure 0.1 Generic project management issues in a construction project.

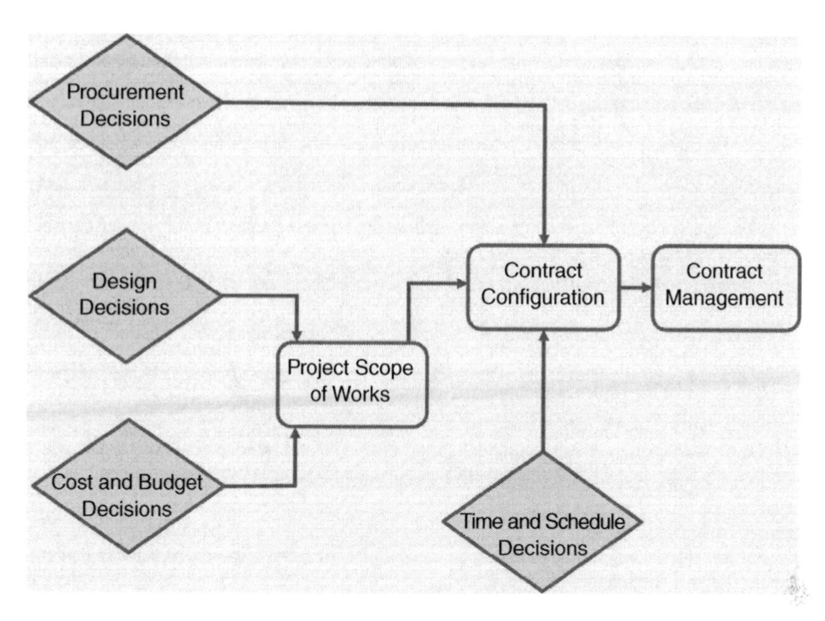

Figure 0.2 Contract administration and management.

(see figures 0.1 and 0.2 above), but have slightly different rules and procedures to deal with the issues, depending on the contract terms and conditions. In this book, I have tried to concentrate on the issues themselves and given general introductory guidance on how to deal with them. More detailed contractual rules and procedures are developed further in more detailed resources. In this way, the reader will be able to apply the general guidance to whichever specific project and form of contract they are working on. The next project will be different, in terms of size, function, cost, location, duration etc. but hopefully the general guidance will still be as appropriate.

Part A

Initiating the construction stage

Contractor mobilisation

The construction stage of a project commences when the contractor is given the Notice to Proceed (NTP) and a date for access to the site. This may seem obvious, but that point can occur at different times in the project life cycle depending on the procurement route which has been chosen for the project.

For example, in the traditional procurement system (figure A.1), the construction stage commences after the completion of both the design and tender stages. The contractor has had little or no input into the design stage and is appointed mainly on price and schedule criteria after the design has been completed by the design consultants.

Alternatively, in the Design-Build system, or other systems which enable early contractor involvement (ECI) – see figure A.2, the construction stage starts well before the design stage has been completed, which allows for some concurrent design and construction, with the intention of shortening the overall project duration and enabling the contractor to input its particular skills to the project design – especially regarding programming, scheduling and buildability.

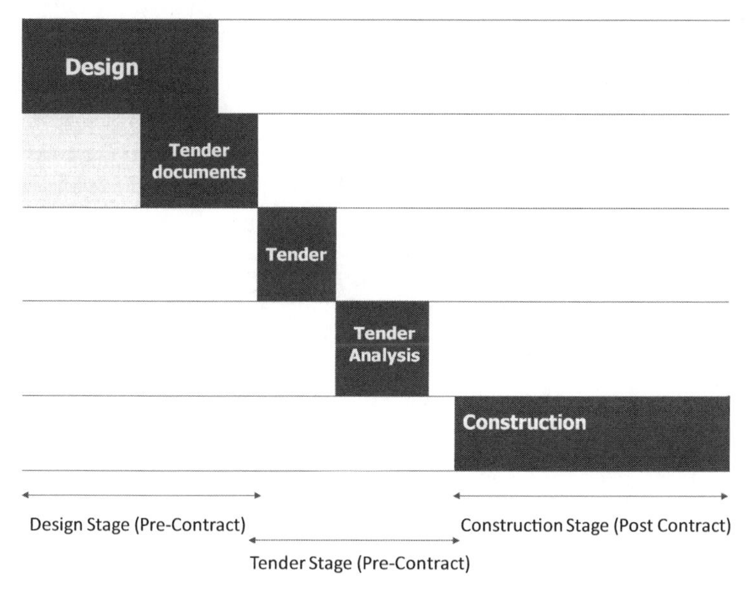

Figure A.1 Traditional procurement of a construction project (design – tender – construction).

This is not a textbook on construction procurement, and the reader is referred to separate texts for a fuller discussion of the relative advantages and disadvantages of the various procurement routes. However, for the purposes of this book, it is important to note that the construction stage can start at different points within the overall project life cycle and therefore the requirements of the contractor and other parties during this stage can vary considerably.

Whichever procurement route is chosen (and only two are discussed above), the contractor will not be allowed access to the site (and for their part, will not commence mobilisation of resources) until the contract has been awarded to them and the Notice to Proceed (NTP) has been issued, although this depends on the particular contract conditions for the project. Under the normal law of contract, the contract is formed when the employer accepts the offer made by the tendering contractor following a period of tender analysis, tender evaluation and possibly post-tender negotiation. The actual award will usually be made by a 'Letter of Award' which will state that a contract has been formed and will set out the full contract documents. In some cases, employers may insist that the full contract documents are agreed and signed before the formal contract is formed.

So, why are we mentioning this here? This book is about the construction stage administration procedures which will clearly be affected by the procurement route chosen in the early stages of the project's life. Procurement decisions affect construction contract administration, as the various project responsibilities are allocated to different parties and all parties need to know what they are responsible for, in order to mobilise the right resources and carry out their duties as efficiently as possible.

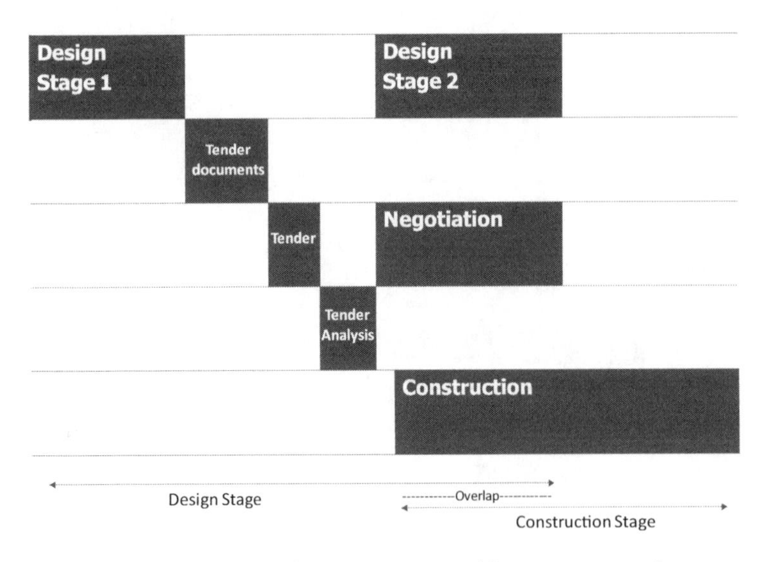

Figure A.2 Two stage tendering / Design-Build procurement of a construction project.

1 Setting up for project delivery

1.1 Introduction

A construction project is an organisation. The difference between a project organisation and a traditional business organisation is that a project is specifically designed to come to an end and finish, whereas a business organisation will – hopefully – continue in perpetuity. The people involved in the project know this and, unless the project is very large, will invariably be separately employed by the different organisations involved in the project.

Put simply, during the construction stage of a project (i.e. the project delivery stage), the organisation set up to actually deliver the facility must be able to concentrate on constructing the scope of work within the agreed time schedule and normally to a cost / budget set by the client. We can therefore adapt the traditional time–cost–quality diagram as shown in figure 1.1.

This chart illustrates that delivering the project scope, within the project schedule and to the project budget (and sometimes requiring a trade-off between the three), is carried out by the project team using project processes which have often been well tried and tested from previous experience and standard conditions of contract which have often been tested by courts of law. Unlike a traditional business organisation, a project is finite – so the scope, schedule and budget have been (or should have been) pre-defined before actual construction starts. This is not a textbook on project management, but it is necessary to understand that managing construction contracts is about creating a team to eventually work itself out of a job, or more accurately, to deliver the project and move on to the next. If the client gets the building or facility on time and within budget, they will be happy and satisfied with the interaction with the construction industry. If the contractor(s) – and any other project participants (such as professional consultants) deliver what the client wants and make a profit into the bargain, they will also be happy and satisfied. If all parties are satisfied with the project outcomes, future business opportunities are invariably easier to come by and negotiate. A good example of this is the Olympic Stadium project in London for the 2012 Games. Clearly, time is of the essence in a project such as this, so the client needs to make sure that the project will be delivered on time (priority 1) as well as fit for purpose (priority 2) and within budget (priority 3). What better way to do that than to employ the entire project team from a recent highly successful delivery of a similar project – namely the Emirates Stadium for Arsenal Football Club?

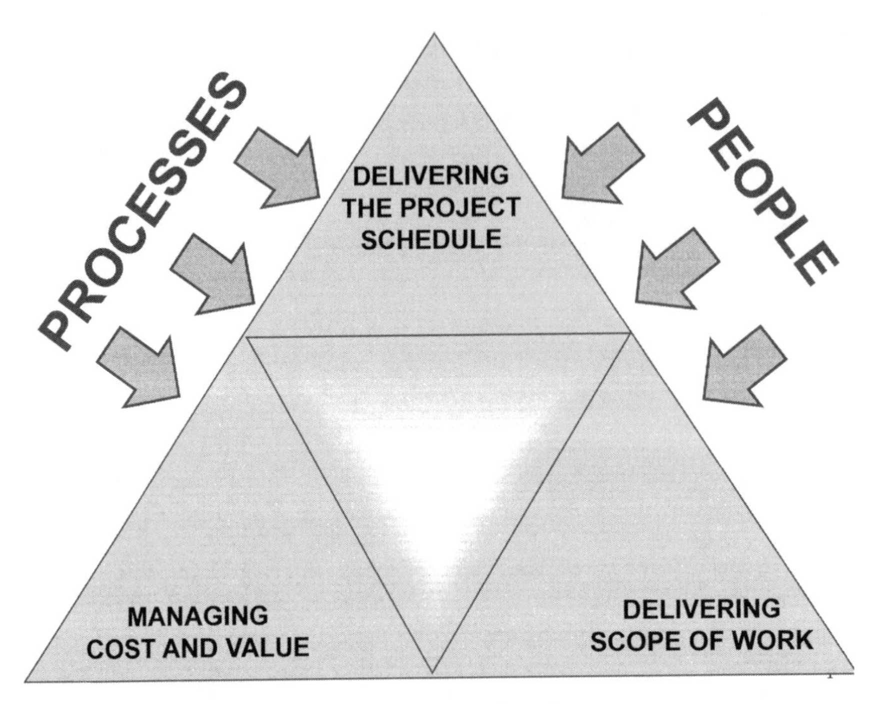

Figure 1.1 Delivering the project scope, schedule and cost.

1.2 Project personnel and procedures

1.2.1 Project personnel

A team can be defined as a group of people with a common purpose who are organised to work together interdependently and / or cooperatively to meet the needs of their customers by accomplishing the goals and objectives. Clearly, in construction projects, the project is a team of people, which is then made up of separate functional teams, all working towards delivering the scope of works, on time and within budget.

For any team to work effectively there must be both clarity and agreement on the expectations of what each team member will do as well as how they do it. This includes the duties, responsibilities and levels of authority of each one of the participants in the project team, which is normally set out within a project management plan (PMP) or project execution plan (PEP) for each individual project being executed. This is especially important for major projects which may take several years to complete and are complex organisations in their own right and with different people likely to undertake the various roles throughout the project's life cycle. As stated above, construction projects are by definition temporary organisations, which are designed and set up at the formulation or inception stage and normally disbanded at the completion of the project when the building or facility is finally handed over to the client for occupation or use. An exception to this would be when there is an ongoing operation and maintenance (O&M) requirements or facilities management (FM) obligation by the contractor, but even then, the facilities management team is normally quite different from the construction team, so there is necessarily some form of hand over.

Additionally, unless the project is very large and requires full time effort by everybody working on the project, many of the people undertaking the various roles will be working on several projects at the same time, usually with different companies undertaking the project functions. As every company has different practices, procedures and cultures, it is clearly of critical importance to develop clear guidelines setting out how each project will be managed and delivered, in order to try to avoid ambiguities, contradictions, misunderstandings and mistakes. All of which can lead to the dreaded D-word – disputes.

For all contractors who tender for construction work, what they want to hear from their directors is 'OK guys, we've got the job now'. This, however, generates different and often contradictory thoughts and emotions – clearly there will be elation that all the work (and cost) which was put in during the estimating and bidding stages will not have been wasted; but on the other hand, if the company got the job by being the lowest tenderer, then what mistakes did we make? How close were we to the next tenderer? During the estimating and tendering stage, the contractor is doing exactly that – estimating what the work is likely to cost and how long each operation is likely to take. These particular chickens only come home to roost when the contract is awarded on a fixed price basis, which means that the contractor now has to deliver the scope of works for the price they have quoted and in the time duration stated in the tender documents. Oh, and hopefully make a profit while doing so.

So, the project therefore needs to be organised properly in order that the client can be confident that they will get the project on time, within budget and to a good quality; and the contractor can be equally confident that they will make a profit on the deal. This all requires confidence that risks and uncertainties have been allocated, assessed, managed and minimised, which in turn requires good management – i.e. planning and control procedures for people, processes, physical resources and costs required on the project.

Let us first look at the roles and responsibilities which are normally allocated in a construction project. Once these have been agreed and defined, the project planning and control will be discussed in Part B of this book. Clearly, there is a great deal of interaction between the two – for example, a design-build contract will have different roles and different project documentation from a more traditional single stage lump sum contract where design and construction are carried out separately but, in the vast majority of projects, there are different roles and responsibilities for those who are acting for the client and those who are acting for / employed by the contractor(s).

Figure 1.2 shows a simplified diagram of the hierarchy of roles on a project and is also intended to illustrate the day to day communications between the peer groups on each side. It is therefore clear to see why there has to be a robust and formal procedure for both communications between the project participants and also document control as letters, memos, drawings, minutes of meetings need to be fully structured and archived for possible later use in case of audit requirements or evidence in any disputes. Chapter 2 takes this issue further in discussing the move towards building information modelling (BIM) as a natural progression and bundling of computer aided design (CAD), coordinated project information and project collaboration software.

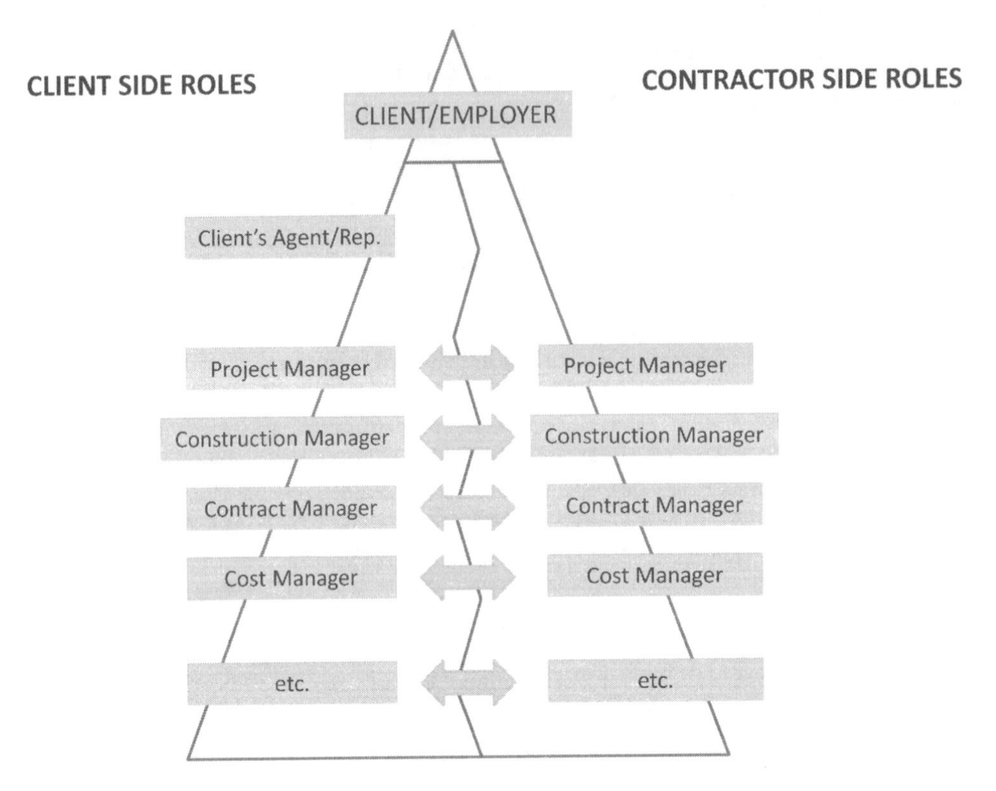

Figure 1.2 Client side and contractor side roles in a construction project.

The personnel / roles required on the client side will include:

The client / employer

It is during the feasibility and design stages that the client should take the greatest care that their objectives for the project are likely to be satisfied by the decisions that are being made. As discussed above, these objectives are normally related to time, cost and quality and a much more comprehensive discussion can be found in the sister book *Introduction to Building Procurement*. When the project has reached the construction stage, the contractor will normally have a detailed scope of work, for which they will be paid the agreed contract price and are required to complete within the contract timescale. During the construction stage, the client must therefore withdraw slightly from the day to day management of the project and leave that to the professional team, although they will obviously retain a keen interest in how the project is progressing. In most projects, during the construction stage, the client's specific functions and responsibilities are to:

- provide the site and allow possession of the site to the contractor during the period of construction;
- appoint professional advisors to carry out functions on their behalf which are essential for the effective and professional administration of the project;
- give appropriate and timely instructions to enable the work to be carried out;

- permit the contractor to carry out the works without interference;
- pay for the work which has been properly carried out and in accordance with the agreed schedules.

As stated above, the client must not hinder the contractor in the progress of the work and must not interfere with any contractual powers of the professional consultants to issue certificates or instructions etc. This invariably does happen on many projects (particularly in international projects) and is the source of many disputes as the client often cannot resist interfering with the duties of the various parties or with the contractor's progress. In some cases, the client considers that they have the authority to issue 'instructions' themselves even though this is outside the normal contractual procedures. In these cases, contractors have a choice – they can either tolerate the interference by accepting it as a cultural norm (and maybe issue a notice requesting a formal Variation to Contract – VTC), or they can object to the interference on strict contractual grounds, thus creating more difficult working relationships, both on this project and possibly in the future.

The client's agent / representative

As most clients are not knowledgeable about the construction industry or its processes and procedures (why should they be, they may well be industrial companies or retail organisations?), then in order to administer the projects, they will appoint an agent or representative to help them or represent them in managing the project and administering the construction contract. This is the professional team mentioned above.

This client's agent or representative may be a person or company whose role is merely to attend meetings and inform the client of any matters which need to be brought to their attention. The client's representative does not normally have any powers themselves, although this will depend on the specific form of contract used on the project. It is interesting to note that under the design-build contract, this role is called the employer's agent but the management forms of contract term it the client's representative. Therefore we have an interesting semantic regarding the difference between an 'agent' and a 'representative'. Under a normal definition, a representative can make decisions without constant recourse to the person appointing them (a member of Parliament is a representative of their constituents and therefore makes the voting decisions themselves), whereas an agent has much more limited powers to merely negotiate on behalf of a client, with the final decisions being taken by the client themselves (e.g. an estate agent).

The project manager

It is becoming common for employers to engage a project manager on more complex projects. This title certainly has more bragging rights but may not, in reality, be more than the client's representative explained above. The project manager role should ideally include managing both the design and construction stages of a project and will therefore cover both the design team and construction team, although in many instances, they are not appointed until the construction stage. Some forms of contract, e.g. the New Engineering Contract (NEC) and most international forms of contract, give the project manager powers to give instructions, take decisions etc. The CIOB Code of Practice for Project Management details the project manager's responsibilities during the construction stage as being:

a to be the proactive 'driver' of the project;
b to set the project objectives;
c to ensure achievement of objectives;
d achieving client satisfaction.

As stated in the Introduction, different standard forms of contract give different names to this role, as the 'engineer' under the FIDIC and ICC Conditions of Contract is effectively a project manager. Therefore, the abbreviation PM / Engineer is used throughout this book to denote the 'leadership' role in a project, whether a particular form of contract uses that term or not.

The design manager

The design manager is a relatively new role within the construction team, especially in the UK where the design–tender–construction system is still the predominant procurement route, so the design is substantially complete before construction starts and the architect or other lead designer has invariably taken on the role of managing the design process. However, where there is overlap between the design stage and construction stage, a design manager will be required to ensure that the deliverables of the design stage are achieved on time. In terms of the design, consultant(s) who produced the original tender or contract drawings and specifications are normally referred to as the designer(s) of record; and they will continue to provide the following design services during the construction stage:

- receive and respond to the construction contractor's requests for information (RFI), communicated from the contractor to the designer through the project manager. An RFI is a request by the contractor for clarification of the design intent of the drawings and specifications;
- review and recommend acceptance of any contractor submittals called for in the drawings and specifications with respect to the construction deliverables;
- review change requests from the original design;
- make periodic visits to the site to assure design compliance (in collaboration with the supervision consultants) and in some cases provide certification that the design has been achieved.

Supervision of the contractor / construction manager / site manager

The construction manager or site manager will be responsible to ensure the project runs according to the schedule or programme and that the works are constructed according to the specifications – i.e. to the required quality of materials and workmanship. Reporting to the project manager, the site manager can either be responsible for a part of the works or the whole site, depending on the size and value of the project. The site manager will be normally responsible for the management of health and safety requirements and to ensure a safe working environment for employees as well as the general public. The CIOB Code of Practice for Project Management states that the role also includes:

- determining how the construction work should be split into packages;
- producing detailed construction schedules;
- determining when packages need to be procured;
- managing the procurement process;
- managing the overall site facilities (access, storage and welfare);
- supervising the package contractors' execution of the works.

Contracts manager

The contracts manager is responsible for the management and administration of the actual contracts within a construction project. This responsibility will primarily focus on the delivery or execution phase of a project, i.e. after the contract has been awarded. A contracts manager may be responsible for overseeing several contracts at the same time and must ensuring that deadlines and budgets within the projects are controlled. The role also includes:

- the management of contractor involvement;
- preparation and negotiation of extensions of time and associated costs;
- ensuring that all insurances, bonds and guarantees are in place and valid;
- ensuring that sub-contracts are managed effectively;
- preparing and negotiating contract amendments and variations as necessary;
- managing the performance of contract administrators.

Cost manager

The cost manager's primary function is to plan, develop and supervise all cost functions on the project, which involves ensuring that all project cost activities such as data collection, estimating, productivity analysis and budget forecasting comply with company and client requirements. This position is responsible for the development and implementation of project cost standards and procedures. The cost manager will implement a standard costing structure to ensure that the project adheres to this standard. The cost department provides a support service to the project teams to assist and guide the establishment of the project budgets and project control tools.

HSE manager

Many employers prefer to appoint a person to take the role of championing their health, safety and environment (HSE) policy and procedures throughout the project life cycle. This role would comprise the following:

- HSE management of the contractor(s), who are expected to have approximately [x] persons on site in multiple locations;
- devising and operating quality control procedures to ensure that the appropriate high standards of construction work are achieved in HSE terms in accordance with client policy;

- undertaking high level risk assessments to identify problem areas for detailed follow-up with chosen contractors (in order to manage identified risk);
- establishing a plan for the reduction of hazardous building practice or materials;
- managing and reducing environmental risk;
- ensuring adequate onsite welfare provision is provided;
- reporting in a regular and timely manner to the client's HSE department on all HSE matters to ensure consistency, continuity and the use of best practice;
- keeping up to date with new initiatives and current world best practice.

Interface manager

An interface on a construction project is any point of connection between different entities working on the project, which could be:

physical	any physical interaction between components;
functional	any differences between functional requirements between systems;
contractual	interactions between client, contractor, consultants, sub-contractors and suppliers;
organisational	information exchanged between disciplines and between parties – see also document control and management in section 2.2;
resources	points of dependencies between equipment, material and labour suppliers.

Because these issues can be very complex and consist of hundreds of different instances, many large projects have a specific interface manager to ensure progress is as smooth as possible. The interface manager will oversee and monitor these interface activities and provide proactive support to nominated focal points within each discipline or organisation. Any particularly critical interfaces will be entered into an interface register and tracked to completion.

On the contractor side:

Contractor's project manager

A project manager employed by the contractor would tend to replicate the role of the client's project manager but with the responsibilities from the 'supply side' of the project rather than the 'demand side', which is the client's main priority. In many projects, the project manager also acts as the construction manager (see below) as both will come from a similar technical background. On larger and more complex projects, where there is a higher need to attend meetings, it would clearly be beneficial to separate the roles.

Contractor's construction manager

This role is mainly responsible for the actual construction carried out on the site and can also be described as site manager, site agent, contracts manager (in UK), building manager or simply construction manager. The role is basically to take responsibility for running and managing the construction site, or that part of it which is under the control of the particular contractor. It is invariably a highly stressful job but very satisfying when everything goes smoothly. The construction manager is a highly regarded and professional position, which can usually only be achieved with substantial experience in the industry.

A construction manager's role typically involves:

- *preparing* the site and liaising with the design team and client's consultants before construction work starts;
- *developing* the programme / schedule of work and strategy for the project;
- *planning* ahead to prevent problems on site before they occur and to make sure that the delivery and storage of equipment and materials occurs in a timely manner;
- *making* safety inspections of the site when work is underway to ensure that all regulations relating to health, safety and the environment (HSE) are being followed;
- *overseeing* the running of several projects at the same time;
- *communicating* with a range of people including the client, sub-contractors, suppliers, the public as well as the workforce on site.

Contractor's contract manager / procurement manager

The contract department of a main contractor will differ considerably from that of a consultant as the main contractor must manage the supply chain for the project, i.e. all the different manufacturers, suppliers and sub-contractors who will supply services, goods and materials to the project. In many cases the contracts manager will deal with the main contract and therefore negotiate with the client and its consultants, whereas the procurement manager will face the other way around and deal with the supply chain as identified above. The role of each or both typically involves:

- setting up a process of review for all contracts, supplies and suppliers to ensure that maximum value for money to the contractor is achieved through supplier rationalisation and by developing a list of approved suppliers;
- encouraging effective contract management across the business with regular reviews, development of service level agreements (SLAs) and key performance indicators (KPIs);
- building and developing good relationships with key suppliers to ensure best value for money and client satisfaction with services provided. This is especially important with regular first tier sub-contractors and suppliers;
- ensuring that all areas of concern are identified and addressed, with remedial action taken as early as possible;
- engaging with benchmarking activities to ensure that value for money and economies of scale are used to the contractor's benefit, where appropriate;
- monitoring trends in the markets for the various materials and services and proposing / implementing plans to respond to such trends;
- ensuring that a central procurement register is maintained and updated.

Contractor's HSE manager

The contractor's HSE manager is responsible for actually carrying out the HSE policies of the company, the client and the project, and must have skills in the following areas:

- HSE leadership and commitment
- HSE policy and strategic objectives
- organising the responsibilities and resources for HSE requirements
- HSE risk management framework
- HEMP for all significant risks (Hazard Effect Management Programme)
- safe systems of work and personal safety
- HSE performance management
- incident classification investigation and reporting
- HSE audits
- HSE management reviews.

1.2.2 Project procedures

On commencement of the construction stage of a project there are certain processes and procedures which will need to be put in place by the project management team, many of which will be virtually identical for every project irrespective of the location and size. A quick glance through the chapter headings of Part B of this book will show what these processes and procedures are designed to do, i.e.:

1. How is the contractor's performance to be supervised to ensure time, schedule and quality compliance?
2. How is the contractor to be paid?
3. What happens if the scope of work needs to be changed?

    a. adding extra items required by the client;
    b. deleting items now not required by the client;
    c. amendments required by *force majeure* – i.e. not within either party's control;
    d. amending the design following value engineering procedures – i.e. better performance at less cost.

4. How to deal with sub-contractors and others who are contributing to the project and therefore require access.
5. How to make sure the project costs remain under control. From the client's point of view, the costs should still remain within the approved overall budget (see figure 1.3) and from the contractor's point of view, that the required profit margin is still being attained.
6. How is the work to be accepted by the client? Will this involve separate approval of materials, goods and services (this procedure is normal in international contracts), or will acceptance and approval only take place at particular milestones during construction?

1.3 The project execution / management plan (PEP / PMP)

The project execution / management plan (PEP / PMP) is the governing document that is developed at the beginning of a project and establishes the means to execute, monitor and control the project. The plan serves as the main communication vehicle to ensure that all parties to a project are aware and knowledgeable about the project objectives and how they will be accomplished. The plan is therefore the primary agreement between all the parties and they should all have signed off

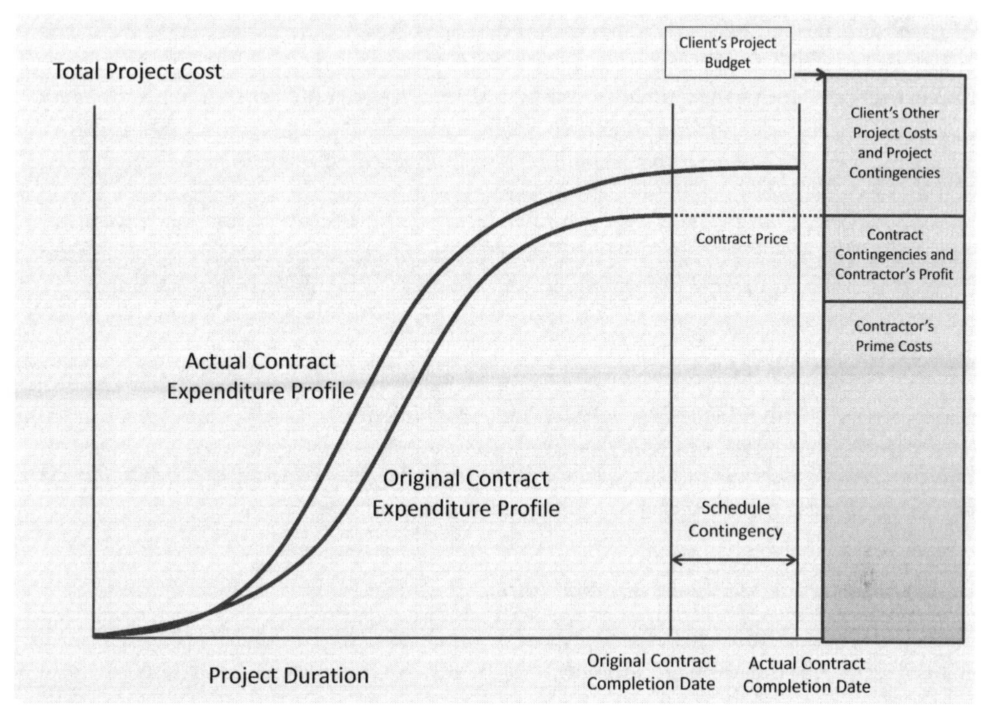

Figure 1.3 Project cost and time.

the document to show this agreement and to 'buy-in' to the proposal. The PEP or PMP should be developed and approved as early as possible in the project and therefore should be well developed by the time that the construction stage is reached. The plan is not a static document and therefore should be regularly updated by the project manager to include current and anticipated future processes and procedures.

Typically, the PEP / PMP will include the following elements:

1.3.1 Project objectives, purpose and priorities

This section will cover the main purpose for which the project is required and include the over-arching project details such as project particulars and key elements from the client's business case where these are necessary to refer back to when changes or variations are contemplated. All too many projects veer away from their intended objectives during the construction stage when a multitude of variations are being applied, so having a baseline statement or cornerstone is critical in keeping the project on track. The business case would include details of funding sources: approved budgets and the project initial programme / schedule (often termed 'baseline schedule'). The defined objectives in this section may not just refer to financial or quantitative benefits as the project may also have non-financial or qualitative benefits, such as community employment, skills development etc.

A project mandate will also be necessary, which includes the initial strategic approval and budget allocation given to the project by the client's board of directors or owners.

This section should also include scope and location details with general arrangements of the physical project together with details of any mandatory or statutory approvals and consents required – such as planning permissions from the appropriate local authorities, building regulations approval etc.

Finally, at the end of the project, how and when will the client know that the project has achieved its objectives? A brief statement of completion date, criteria for successful completion and requirements for occupancy should also be made in this section.

1.3.2 Critical risk analysis

All projects which involve future activity have an element of risk associated with them. These risks should have been analysed and assessed before giving the go-ahead and this analysis / assessment should be highlighted in the plan. See section 3.2 in Chapter 3 for further discussion on risk assessment and evaluation.

1.3.3 Organisation, roles and responsibilities

The PEP / PMP should describe how the project is to be organised, together with all necessary institutional arrangements for the various project participants and including:

- the names, addresses and contact details of all organisations involved in the project, from both the client side and contractor side including stakeholders and subcontractors. The hierarchy of authority should be clearly set out;
- their roles and relationships with each other, preferably in graphical format;
- their particular responsibilities and the delegated level of authority;
- details of any other stakeholders not mentioned above.

In particular, the roles, responsibilities, accountabilities, delegated financial authority for the main participants in the project plus first tier contractors will be set out in detail in the plan. Good project management requires the allocation of single point responsibility for each work component or element in the work breakdown structure. In this way, nobody can avoid or deflect responsibility for non-performance.

1.3.4 Project strategy

The plan should set out how the project will be managed and controlled in terms of design, procurement and construction.

Design

- identify the design deliverables and key design milestones (i.e. at each stage of the design process);
- Health and safety requirements during the design stage (e.g. HEMP – Hazard Evaluation and Management Programme);

- procedure for controlling changes to designs and authority for approving changes – see also section 6.3 in Chapter 6;
- arrangements for design reviews, whether formal or informal;
- design coordination responsibilities. this element should *never* be underestimated, especially where there are multiple design companies involved in a project – e.g. architecture, structural engineering, MEP (mechanical, electrical and plumbing) services, interior design etc. On far too many projects, the holes through walls, beams and columns provided by the builders are in a completely different place to where the cables and ducting will go. Don't laugh, it happens;
- planning permission, submission to the appropriate authorities and time taken for approvals.

Procurement

The plan should set out the decisions related to the procurement of the project, for both materials and services. Needless to say, the procurement route chosen should be the most appropriate for the requirements of the project and the choice of tendering procedures and contract documents will follow from the choice of procurement route. Many client organisations operate a system of 'procurement strategy workshops' in which all the internal stakeholders plus selected external parties decide on how to most effectively procure the facility. For example, should the design be completed in total before contractors are appointed, or would the project benefit from the early involvement of contractors for their construction and programming expertise? These are decisions outside the scope of this book, but once the decision is made, the actual contractual relationships and contract documents will flow from this procurement decision.

Construction

The project strategy for the construction stage together with objectives, programme / schedule, procedures and control systems established at the pre-construction stages also need to be included in the plan, including:

- site logistics, constraints and phasing, including health and safety issues;
- milestones and coordination of sub-contractors' programmes;
- contract administration roles and responsibilities;
- permits, site security and safety;
- interfaces with utilities and statutory bodies, such as electricity company, water company etc.;
- insurances of the works: who is responsible?
- bonds, guarantees and warranties, as required.

1.3.5 Project controls

The PEP / PMP should also include the agreed control mechanisms for the following construction stage issues, which will also be included in the conditions of contract. The project manager should clearly ensure that what is included in this section of the PEP is strictly in accordance with the conditions of contract and other formal contract documents of the project:

- *Changes / variations:* all construction projects in the history of the world have been subject to change of some sort during the construction stage, i.e. after the contractor has been appointed with a particular scope of work. Therefore, and especially if the contract is a lump sum, unless there is a contractual procedure to formally vary the works (i.e. add / omit from the scope of works or extend the completion date), any amendment issued by the client is technically a breach of contract. The PEP / PMP would include procedures for initiation of change requests, preparation of change proposals, authority for and approval of changes, registering project changes within the project scope of works – termed configuration management.
- *Programme / schedule*: which may or may not be a contract document. Usually the initial programme (sometimes referred to as the BLS – baseline schedule) is not a contract document, as this has particular legal ramifications, especially when variations to it are considered during the construction stage. The plan would include the outline programme comprising the work breakdown structure (WBS), key milestones and activities with the main resources allocated. There may also be an indication of areas of potential slippage of the schedule (in order to highlight these areas to key decision makers) as well as key decision dates in the project. At this stage, a simple Gantt chart will be sufficient for inclusion into the PEP / PMP.
- *Cost and value*: the actual agreed cost / value of the contract, plus the formal mechanisms for amending the cost due to variations or other authorised reasons for changing the contract value. The payment mechanisms to the contractor and other project members should also be explained, whether based on milestones, measured work or cost plus. Additionally, the procedure for regular cost reporting will be included, which is intended to give an early warning to all parties of potential project cost overruns or possible schedule delays.
- *Quality*: given that the project delivery is a transaction – the contractor builds the project and the client pays them money, the contractor's performance needs to be supervised and the final product formally accepted. The 'quality' should be defined in the contract documents – usually the project specification and the procedures for assessing the quality will be explained and outlined in the plan. This will necessarily include administrative systems for materials approval, notices by the contractor for major operations such as concrete pours, Inspection Certificates etc. together with mechanisms for rectifying defects etc.
- *Organisation and administration*: as stated previously, the project needs to be organised both professionally and effectively so that all parties can achieve their objectives. The organisation needs to be clear and unambiguous with little room for uncertainties. On large projects, there is often a separate 'document control' department on both the client / consultant side and also the contractor side, to ensure single point transmission and archiving of project documents (letters, memos, drawings etc.). See also section 2.2 in Chapter 2 for document control and management. Regular project meetings will necessarily have to take place throughout the construction stage and the PEP / PMP will set out the standard operating procedures for these meetings, including objectives, frequency and required attendance.
- *Risk*: Uncertainty and risk are synonymous and risk also creates hazards which can in turn lead to accidents and lost time incidents (called LTIs in health and safety jargon). Therefore, the effective management of risk is critical to the effective management of construction projects.

- *Progress reporting*: effective control relies on comparing what has been done with what was expected to be done. Therefore, reporting on actual progress against the original programme / baseline schedule will show if progress is ahead of schedule, on schedule or behind schedule. In each case, the strategy for remedial action should be included in the PEP / PMP. Frequency and responsibility of both cost and schedule reporting should be set out in the plan.
- *Key performance indicators (KPIs) and critical success factors (CSFs)*: When monitoring progress, the project manager will necessarily concentrate on the KPIs and CSFs in ensuring the project is being delivered both effectively and efficiently. The KPIs and CSFs should be set out at the beginning of the project for the awareness of all stakeholders, which will act as the high level targets for the project manager.

1.3.6 Commissioning, operation and maintenance

Following completion of the construction operations and when the building or facility is put into use by the client or end-user, the contractor will still have an involvement during the defects liability / defects correction period. Additionally, the project may include some commissioning or operation of the facility by the original contractor, which is more common in process engineering projects, such as oil and gas facilities. Therefore, the PEP / PMP would include some specific technical standards for commissioning, operation and maintenance not covered in the construction brief, for example:

- *User commissioning and acceptance testing*: for any specialised equipment – standard MEP installations will normally still be the responsibility of the contractor.
- *Handover meeting / procedures*: for both practical / substantial completion and final completion.
- *Format of as-built or other permanent record drawings*: most projects nowadays require the contractor to hand over a complete set of as-built drawings, especially on those projects where the contractor is responsible for the detailed final design. The plan will consequently include details of how these documents should be delivered.
- *Operation & maintenance (O&M) requirements*: relating to the O&M instructions for plant and equipment, including plant and equipment numbering system to be compatible with the asset register used by the client or end-user.

1.4 The Pre-start / Kick-off Meeting

This is a meeting which should be arranged by the project manager before any construction or contractor mobilisation commences and is intended to 'kick-off' the project by setting the scene for all future communications and working relationships. Although this football analogy is useful, the meeting is probably a closer equivalent to the line-up and team handshake, since once the football match has kicked-off, the game is in full swing, which is not necessarily the case with the Pre-start meeting for construction projects. Interestingly, there has never been any suggestion of using the equivalent term in ice / field hockey (bully-off). This may be far more appropriate given the behaviours of some parties to construction projects.

There are several benefits to site communication which may arise from such a meeting, including:

- It allows people to get to know each other, which is likely to lead to better project communication and less confrontational attitudes as the work progresses.
- It provides the opportunity to decide on how the project communication systems will operate.
- It provides the opportunity to define points of contact (focal points) within each organisation.
- It enables all relevant contact details to be shared for team leaders working on the project.

The Pre-start / Kick-off Meeting is critical to the success of the construction phase as it establishes and ensures all parties are aligned with the construction phase procedures, key milestones, areas requiring special interest, communications, submittal requirements, any required amendments to the contract documents as well as the goals and objectives for each of the parties to the project. A well-prepared and successful Kick-off Meeting will have a major impact on the subsequent success of the construction phase.

1.4.1 Pre-start / Kick-off Meeting agenda

All formal meetings must have a well-prepared agenda if they are to be effective, which should be distributed to the invited members well in advance, together with any information or documents which are intended to be read in advance of the meeting (often termed 'pre-read'). Some of the agenda items are for the contractor to prepare and present and others will be for the client or their consultants to prepare and present. In all cases, the size and complexity of the project will determine the points to be covered in a Kick-off Meeting and the length of discussion for that item.

A standard agenda for a Pre-start / Kick-off Meeting would include:

1. Election of chairman (normally the client side project manager)
2. Introduction of attendees
3. Distribution of construction procedures manual or contract / project execution plan (CEP / PEP)
4. Design scope, including finalisation of design and requirements for shop drawings etc.
5. Lines of formal project communications
6. Construction stage meetings, including frequency, location and chair / secretary
7. Site offices – location and size
8. Site set-up, including perimeter hoardings and signage, special security arrangements
9. Safety issues, including training, medical facilities etc.
10. Construction contract breakdown (work breakdown structure – WBS), including cash flow
11. List of sub-contractors, including requirements for approval / registration
12. Permits and licenses required
13. Project schedules and programmes, including procedure for approval / endorsement
14. Ensure that all insurances, warranties, bonds and guarantees are in place
15. Testing and inspection requirements for both client and contractor
16. Any other special requirements, such as client use of site, client-issued materials, other directly employed contractors, phased works, sectional completion, partial possession etc.

17 Procedures for construction progress reporting including clarifications and possible changes
18 Procedures for contractor submittals, including, shop drawings, samples, mock-ups, materials etc.
19 Procedure for disputes – including contractual notices, early warnings etc.
20 Procedure for taking over the works, including partial possessions or sectional completions.

Clearly, getting through this entire agenda will take some considerable time, therefore it is important that issues have been prepared well in advance with the necessary documentation, which can be tabled at the meeting and subsequently archived as a record for future reference.

1.5 Contractor's mobilisation

The beginning of the contract is a period during which both the contractor and employer each have several obligations to fulfil including:

This early period in the contract is often mentioned by the contractor in subsequent claims for extensions of time, for example, if the employer has not fulfilled an obligation (such as late or problematic possession of the site). The PM should also consider whether the contractor has contributed to the delay by a failure to fulfill their own obligations, i.e. being sufficiently mobilised to perform the works.

It is therefore crucial that the project management staff keep detailed daily records of the contractor's mobilisation activities on site until they are fully mobilised to perform the works, which will include:

- dates of mobilisations of contractor's representative (senior staff) and numbers / categories of contractor's personnel (incl. sub-contractors);
- dates of mobilisation of numbers / categories of contractor's plant and equipment and any necessary materials for temporary works;

Table 1.1 Project obligations of the contractor and employer

Contractor	*Employer*
• Appointment of contractor's representative	• Appointment of PM / Engineer (depending on title in the contract)
• Submissions of or provisions for provision of drawings	• Mobilisation of supervision personnel?
• Appointment of contractor's representative	• Appointment of [PM]
• Submission of:	• Mobilisation of [PM] personnel
– Performance security	• Provision of design Information
– Advance payment security	• Acceptance of:
– Insurances	– Performance security
– Lump sum breakdown and unit rates	– Advance payment security
– Work programme / schedule	– Insurances
– General mobilisation	• Giving access to and possession of site
	• Obtaining statutory approvals and permits

- dates of delivery of materials and plant;
- progressive dates of contractor's site installations (offices, yards, lay-down areas, laboratories etc.);
- progressive dates of contractor's installations and provisions for supervision by the engineer / PM;
- dates of necessary statutory approvals and permits (if relevant).

1.6 International aspects

Clearly then, there is plenty to do at the beginning of the construction stage, when the contractor is being mobilised to start the works in earnest. For international contracts, there are several extra dimensions for the client, project managers and contractors to consider. If the project is located, say, in the Middle East, the client will undoubtedly be an organisation established in that country and will have arranged funding for the project in the local currency – rials, dirhams or dinars. The project manager and other consultants may be local companies, but will most likely be international specialists operating from a local office. The contractor may also be a local company and there are now many Middle East based contractors who are fully capable of delivering major projects, although most will still employ expatriate specialists as well as expatriate labourers, semi-skilled and skilled operatives. Specific international aspects will include the following.

1.6.1 Legal jurisdictions

All contracts are agreed under a particular legal jurisdiction. For a contract in London, then English law will apply and the terms and conditions must satisfy the requirements of English contract law. Dispute resolution procedures will be covered by the 'Construction Act', arbitration clauses will need to satisfy the Arbitration Acts, health and safety is covered by the Construction (Design and Management) (CDM) Regulations etc. The English legal system has jurisdiction over the contract and as the legal system is very mature, there are many terms and conditions which would be implied into construction contracts, whether or not they are explicitly stated in the contract documents between the parties.

A project which takes place in, say, the United Arab Emirates (UAE), will similarly be regulated by the national laws and courts of the UAE. There may not be equivalent legislation to cover alternative dispute resolution (ADR) procedures, although there is extensive legislation covering arbitration. In all cases, the project is under the jurisdiction of the UAE commercial court system, which is based on Islamic or Sharia law and therefore operates differently from the western legal system.

It is possible for the client and contractor to agree in the contract documents that the contract will be based on the laws of a legal jurisdiction different from the location of the project: for example the contract could state that the terms and conditions are based on English law, even though the project takes place in, say, Qatar. Great care must be taken in these situations and local legal advice is required to ensure that the local legal jurisdiction can be by-passed in this way.

1.6.2 Language of the project

Clients, consultants and contractors from different countries and cultures will invariably have different primary languages; therefore the vast majority of international project documentation will be in the English language, which has become the *lingua franca* of the international construction industry as well as much international business overall. This, of course, is mightily convenient for those of us whose primary language is English but there is a danger that natural English speakers expect others to understand concepts in the same way that they do, which may not be the case. Even within the English speaking world, the understanding of terminology and definitions may be different – for example, to an American, the term 'design and build' refers to the procurement system whereby design is separate from the construction, i.e. design AND build. In the UK, this term refers to the single point responsibility for both design and build. The Americans would refer to this as design-build. During project meetings and in project correspondence, it is always advisable to use plain language and keep it simple.

1.6.3 International tendering

Tendering for major international projects may mean that the estimating department of the tendering contractors is located in a completely different country from the project location and client / consultant offices. There will be time zone differences as well as weekend differences (Friday and Saturday is the normal weekend in Muslim countries) thereby reducing the possibility of direct communication. The risks which the tendering contractors need to mitigate can be very different – ranging from economic risk of the country to physical risk of the climate.

1.6.4 Multiple currencies

One of the risks that the contractor will need to assess is that of exchange rate fluctuations from the currency of the project to the currency of their own home country. Where the project takes place in a country with a freely convertible 'hard' currency, the payments will often be made in that currency – so contractors in Europe will be paid in euros and contractors in the middle east will be paid in rials, dirhams, dinars etc. Some clients, for example in the oil and gas industry, operate in US dollars because that is how they are paid on the international oil markets; therefore, all their major capital expenditures are also based in US dollars. In either case, it is the contractor's risk when converting to their own home currency. Under the NEC contract, the employer may accept the exchange rate risk as a 'secondary option'. In countries which do not have freely convertible currencies, contractors will be paid a proportion in a 'hard' currency (usually US dollars) and a smaller proportion in the local currency, for everyday payments in the country of the project.

1.6.5 International mobilisation

It can often take several months for an international contractor to fully mobilise for a project in a different country. In terms of people, many countries now insist on a proportion of workers being locally employed (see section 1.6.6 below), therefore there will need to be a period of time for advertisement, interviewing, training etc. With expatriate staff, accommodation will need to be provided, and client approval and working visas obtained before any project specific training can be given. In terms of plant and equipment, the early stages of a project usually involve excavation and earthworks, piling etc. which can require large and expensive items of plant to be transported to the country and to site. This is an expensive period of time for the contractor, who is unlikely to receive any project payments until work has been carried out, measured and processed. For this reason, many international forms of contract include a provision for an advance payment to the contractor, which is repaid over time as the project is being built. In order to protect the client, the contractor is often required to provide an advance payment bond which will repay the client should the contractor default.

1.6.6 Increasing local content and in-country value (ICV)

Many developing countries are looking very closely at their capital expenditures and the amounts which are retained in the country as opposed to overseas payments to foreign companies and suppliers. This has become even more important in countries where a high proportion of the local population is unemployed and can see that the international contractors are bringing expatriate labour from abroad to work on the projects – at all levels, from unskilled labourers to managerial staff.

This local content initiative, whilst clearly being worthwhile from the host country's perspective, also means that the tendering contractors will not be able to resource the projects as they would wish with the most cost effective resources at their disposal and are also required to submit themselves to a rigorous prequalification or validation procedure and possible post-contract monitoring procedure, which adds even more costs to their company overheads.

It is likely that, in the future, more developing countries will engage with this initiative in order to speed up the overall development of their own economies.

1.7 Summary and tutorial questions

1.7.1 Summary

The commencement of a construction contract is a period during which both the contractor and employer have several obligations to fulfil. These include all of the issues itemised in table 1.1.

Fortunately, most experienced construction companies are used to this 'start-up' procedure as it clearly needs to take place on every project with which they are involved. All projects should therefore follow the same procedures of kick-off, start-up and mobilisation and the procedures used should form part of the QA documentation of all the project participants and be well documented so that all parties are familiar with the decisions made at this crucial time.

As the management procedures are almost identical for each project, irrespective of the type of project, the roles that are carried out are similarly identical. The 'client side' roles described in section 1.2 are mirrored by the contractor side roles and each role needs to work and communicate with its equivalent to ensure effective management of the project.

At the project inception, a project execution plan (PEP) or project management plan (PMP) should be prepared and published the show how the project will be executed and managed. All subsequent actions must be able to be related back to the contents of this document.

This early period is often cited by the contractor, in any subsequent claims for extension of time, claiming that the employer has not fulfilled an obligation, e.g. late or restricted possession of the site. Therefore, the PM / engineer needs to be aware of this and assess whether the contractor was indeed held up or just using this as an excuse because they were not fully ready to commence work themselves.

1.7.2 Tutorial questions

1. Discuss the tensions in the time–cost–quality triangle.
2. Outline the essential differences between a client's agent / client's representative and a project manager.
3. What are the core skills required of a contracts manager and / or a cost manager?
4. Draft and summarise the contents of a project execution plan / project management plan.
5. What is meant by 'mobilisation'?
6. Discuss the purpose of a Pre-start / Kick-off Meeting.
7. What is meant by 'schedule contingency' in figure 1.3?
8. What are the additional risks to a contractor of working internationally?

2 Project communications, document control and BIM

2.1 Project communications

Communication is the process by which information is encoded and transmitted by a sender to a receiver via a channel or medium. The receiver then decodes the message and will often give feedback to the sender. Therefore, communication is a process where the sender tries to convey a message or a meaning to a receiver and, critically, hopes that the receiver gets the same meaning. If not – that's a communication breakdown, which is how the vast majority of disputes originate – which is why it is critical to discuss the issue at this point. See figure 2.1 for a graphical chart of this process.

Most businesses depend on communications in order to function, since the output from one party (for example, the architect's design) forms the input for another party (i.e. the contractor). Project communication and management tools and techniques are designed to (hopefully) ensure the effective, timely and appropriate generation, collection, dissemination and ultimately disposal of project information. Therefore, communications planning, flow, structure and standardisation in a project organisation are critical in ensuring successful project delivery.

Proper and effective communication skills are essential at all stages of a project from inception to completion together with document standardisation in terms of titles, labels, reference numbers and even size of paper, which can complement those skills by ensuring that all project documents maintain a coherent structure and are more easily referenced, archived and retrieved when required.

2.1.1 Written communication

In a construction project, written communications would clearly include letters, emails, faxes, memos, reports and other project documents in both hard copy and soft copy. All communications and documents should follow a formal document control and transmittal procedure to note when and by whom they were generated, sent and received. In some legal jurisdictions, emails are not accepted as formal written evidence, so it is the responsibility of the project management team to ensure that all written communication is legally acceptable, if required. This is the main reason why it is essential to use a robust project collaboration software package which allows all such documents to be transmitted and archived with a full audit trail. The systems also allow various levels of permissions for project members to access certain documents. See also section 2.2 below.

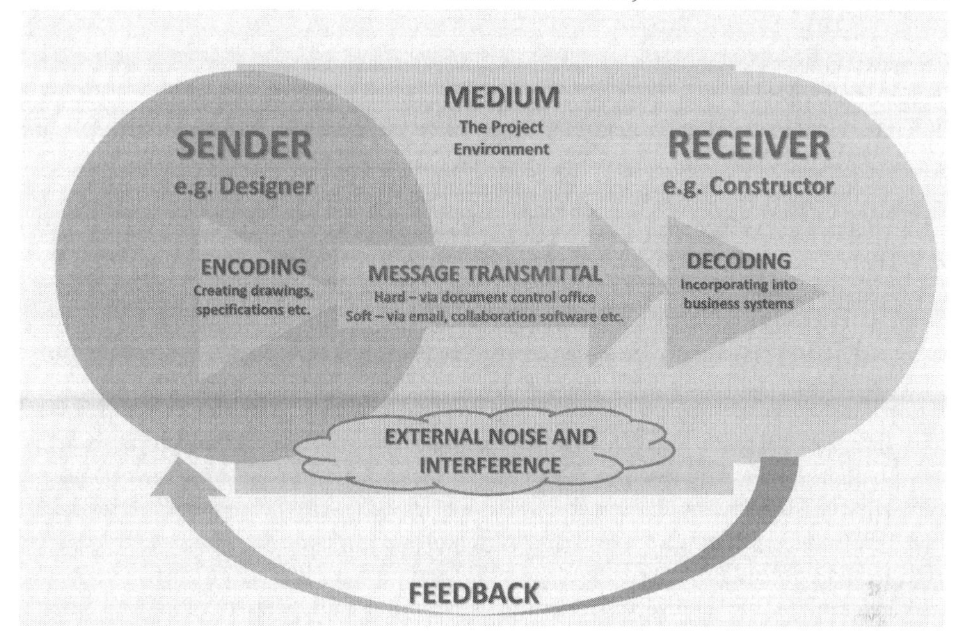

Figure 2.1 Standard model of communications.

2.1.2 Verbal communication

In a construction project, verbal communications would include discussions, presentations, telephone calls, voicemails and even informal chats in the corridor or at the coffee machine, if the receiver is relying on that information as part of their project duties. The obvious difficulty with verbal communications is evidence – proving that the communication took place and what was said. This is why all project meetings should have formal minutes with a protocol for subsequently agreeing the minutes so that they become a true record of what was said and / or agreed at the meeting. Presentations would be normally accompanied by a PowerPoint (or similar) presentation, which effectively converts the verbal to a written communication and will include all the relevant points given in the verbal presentation.

2.2 Document control and management

All construction projects generate vast quantities of documents, ranging from drawings, specifications, pricing information, programmes, schedules, reports, letters, memos etc. It is essential that all of these documents are stored in a secure environment and capable of being retrieved by authorised personnel as quickly and efficiently as possible as well as being revised, updated and added to as the project work progresses.

2.2.1 Document control

The essential features of a good document control and management system are:

a security and access control
b file management
c version control
d ability to view documents quickly
e interface with project communications tools
f output document in a format readable by standard office software.

Construction project teams can, and do, consist of hundreds of people all simultaneously using thousands of separate documents and exchanging millions of individual items of information and data. Therefore, the time spent by each person in generating, distributing, reviewing, approving and retrieving this information can have a major impact on project schedules as well as their employer's profitability. This information can also be subject to complex processes if multiple organisations are involved in the project – which of course they invariably are. Therefore, in order to ensure project efficiency and effectiveness, it is in everybody's interests for these processes to be managed on a single secure and robust platform, to help capture and store the information more comprehensively, thus reducing project complexity and increasing control. There are several proprietary software packages on the international market which will perform this function and when choosing a document control and management package, it is necessary to ensure that it is:

- secure and reliable, with robust well-tested technology;
- easy to use, for both technical and non-technical staff and preferably requiring no complex software installation on desktops, laptops or tablets, which can just be one more thing to go wrong!
- searchable: an internal search engine will help project members find required documents and correspondence easily and quickly;
- designed and structured according to industry standard processes with real time version control so that the latest version of a document is always on top;
- fully auditable from document revisions, to approvals and request for information (RFI) responses etc., promoting adherence to project processes and thereby reducing the chance of disagreements and disputes. This also provides a clearer path to resolution should disputes occur in the future;
- easy to generate project reports: especially regular weekly or monthly reports using (or easily exportable to) standard shell software applications such as Microsoft Word or Excel.

Other requirements of good document control and management are:

Traceability

All document revisions must be logged, so that it is always clear which documents have been sent where and when. The document logs must be visible and also show, if possible, which document revisions are late together with the party responsible. A very useful technique which is incorporated into many proprietary systems is colour coding or a traffic light system to show document status reminder emails to responsible parties.

Intelligent numbering

Given the opportunity, every organisation would have their own document numbering or referencing system, which would clearly be a problem in itself when working collaboratively with other organisations. The numbering system should be clear to all participants as well as capable of automatically generating numbers for new documents as the project progresses.

Standard reporting templates

Standard reports, generated at either regular time intervals (weekly, monthly) or at specific milestones (completion of elements or sub-elements) should be a feature of an intelligent document control and management system. These standard reports may vary from straightforward listings to status reports, late reports, reports grouped in terms of the responsible engineer, status code, approval code, recipients etc. An automatic ability to produces graphs, bar charts and line charts within the reports is also a major advantage, although most standard spreadsheet packages have this facility.

Document tracking and distribution

This can be carried out by using a 'document distribution matrix' where combinations of metadata such as discipline, document type, status code etc. can indicate to whom the documents need to be sent and also log when the document has been sent, when it was downloaded and opened, what changes were made and when it was subsequently forwarded on. It is vitally important to keep control of document tracking and distribution, so the distribution matrix can also act as the transmittal form from one person to another.

Revision control

Clearly, only certain people within a project will have the authority to revise documents, especially documents relating to the scope of works, such as drawings and specifications. Therefore, revision control is paramount. The document control system must have the ability to store all document revisions but only distribute the latest revision (unless a previous revision is specifically requested). The latest revision must therefore be shown by default (always on top) giving the revision number and date of revision with appropriate authority.

Reviewing and marking up documents

Design engineers should be able to review and mark up project documents (sometimes called 'redlining'). Therefore the document management system should have a facility to register comments and upload comment files and preferably for more than one person to work on a document simultaneously, to avoid delays. Engineering design managers may also need to merge and approve comments and revisions from the design team. When a review task has been completed, the next person in the workflow should be automatically notified.

2.2.2 Document management

Modern technology allows information and data to be stored in 'the cloud' and then retrieved securely by those who are authorised to do so. Most people also now use hand held electronic devices such as tablets and smartphones to access work related data, which themselves have limited storage capacity. Therefore document management systems need to accept that documents are stored somewhere in this nebulous 'cloud' and require web-based interfaces in order that the documents can be accessed from anywhere providing the user is connected to the internet. The document management system should have the facility to check out and check in particular documents, specific user inboxes as well as clearly needing a very secure central repository. Through the web-based interface, the user community, both internal and external to the project, should be able to directly access the latest document revisions and status information, correspondence items and technical queries. Searching and reporting should be supported together with content search and drill down facilities to allow downloading to standard office software for word processing and spreadsheet applications.

Therefore, the main components of an effective document management system (DMS) should be:

a central repository with global accessibility
b drill down and search capability
c check out and check in facility
d interface with standard office software (word processing, spreadsheets etc.)
e integrity of data
f workflow integration
g interface with third party document management systems.

Central repository with global accessibility

As mentioned above, project data can be made available from any location in the world when stored in the 'cloud'. Therefore, at any time and from any location the project stakeholders will have direct access to the latest revisions and be able to monitor project progress. Documents, drawings, correspondence items and technical queries are stored in the central repository.

Drill down and search capability

Good document management system should have multiple drill down and search options with different search trees available as standard. Searching down the tree should find the required document revisions based on metadata. Direct attribute search (e.g. document number, title, creation date) and content search should also be supported. Records can be found via the planning tree or the asset item tree. End-users should also be able to build their own folder structures, adding documents from the central repository and monitor progress per folder.

Check-out and check-in facility

When a document revision is checked out by an authorised user, no other person can work on the document revision as it is temporarily reserved for editing. After the document is checked back in again a notification of the update will be generated. Clearly, the person who has checked out the document needs to complete their work in the shortest possible time, especially for documents which others also need to use.

Interface with standard office software

DMS needs to be able to store all correspondence items including emails, which can be migrated from the company or project's email software and include information such as sender / recipient details as well as subject and message.

Integrity of data

All data input needs to be verified against project reference lists so that no incorrect information can be loaded into the system. Document numbers can be checked for correctness of the numbering and correctness of the different elements. This also makes document retrieval faster and easier.

Workflow integration

Workflows should be relatively easy to set up and should both support and streamline the organisation's processes and procedures. Progress of the individual items through the workflow can be monitored by the management and every user involved can find their own outstanding workflow actions in a personal inbox.

Interface with third party DMSs

As different companies may use different DMS packages, it is clearly beneficial for them to be able to talk to each other. Uploads into and downloads from each DMS should be fully transparent, meaning the user does not even notice that the files are stored in a different file management system. All interfaces should support all types of data manipulation, including check out and check in.

Project and corporate documents are normally structured in different levels, from the general higher level documents such as corporate manuals, mission statements etc. to the detailed and specific documents used on a daily basis. Figure 2.2 shows this hierarchy of corporate / project documents which must all be incorporated into the DMS with all the requirements mentioned above. The main issue for a project-based DMS is to ensure effective interface between level 3 and level 4 documents between all the project participants.

2.3 Contract documents

No discussion of project communications and documents would be complete without a section on the actual contract documents on a construction project, as these are the documents that form the basis of the agreement between the parties and their contents are therefore legally enforceable.

Construction 'Contract Documents' are defined as the documents which comprise the construction contract. Right, well that's fairly self-evident, so what are the normal contract documents in a construction contract and what does that mean?

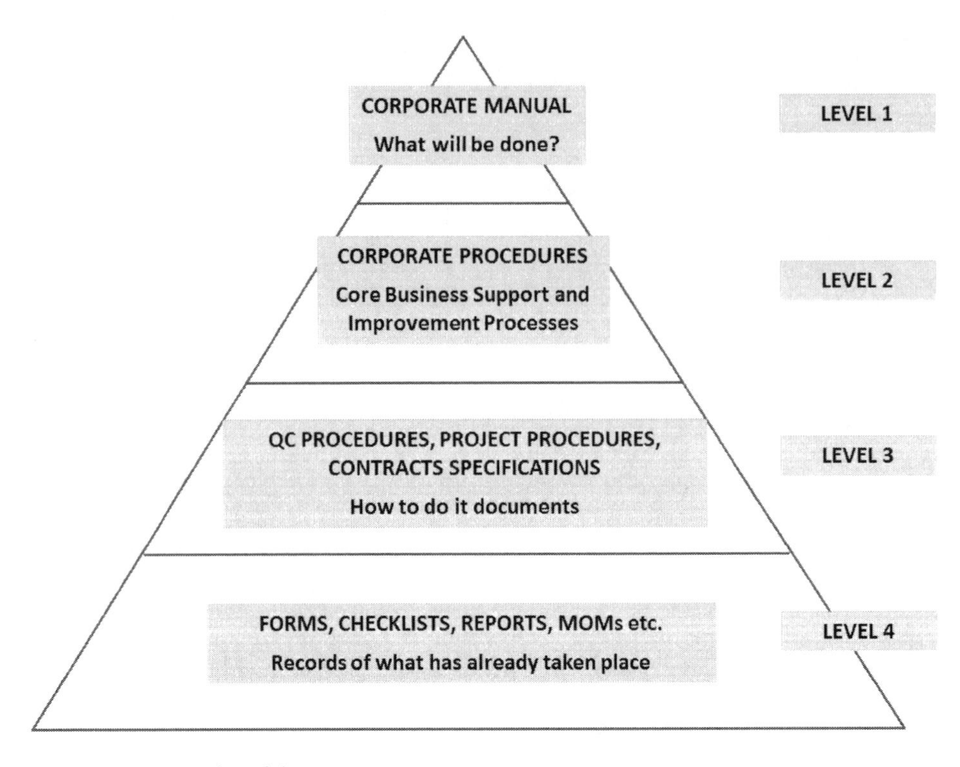

Figure 2.2 Hierarchy of documents.

Under UK law – i.e. the 'Construction Act', all construction contracts must be in writing. In other jurisdictions, this may not be the case but the parties would be strongly advised to ensure that the contract between them is in writing, so that there is no confusion or ambiguity of understanding. The contract documents are therefore the written documents that define the roles, responsibilities and scope of work under the construction contract, and are legally binding on both parties (i.e. employer and contractor). The individual documents that constitute the set of 'contract documents' will be defined in the contract agreement and may change depending on the route chosen as well as the particular standard form of contract used.

On a traditional, fully designed project, the contract documents will normally include:

- Articles of Agreement together with the Conditions of Contract;
- contract drawings;
- Bills of Quantities;
- specifications;
- information release schedules, if all design information has not yet been made available to the contractor;
- a list of tender adjustments or clarifications negotiated and agreed after the receipt of tenders and prior to the signing of the contract;
- the requirement for the contractor to provide any required bonds, warranties and guarantees.

It is best practice in contract administration for both the employer and contractor to sign the contract prior to commencement of work, although in many cases the desire to begin construction means that work has invariably started before the contract is formally executed. This is not a major problem if a letter of award has already been issued and relationships are cordial (which they usually are at the start of a project). However, in such circumstances it becomes more difficult to resolve disputes, so the earlier the contract documents are formally defined and adopted, the better for both parties.

On design and build projects, the contract documents will be different as the full design is not yet developed at contract execution stage, therefore there can be no Bills of Quantities either. In this case, the contract documents may comprise:

- the articles of agreement and conditions of contract
- the employer's requirements
- the contractor's proposals
- the contract sum analysis.

On projects that adopt Building Information Modelling (BIM) – see section 2.4 – the contract documents may also include:

- a model enabling amendment introducing a BIM protocol as part of the contract documents;
- a BIM protocol, which establishes specific obligations, liabilities and limitations on the use of building information models and can be used by employers to cover particular working practices;
- employer's information requirements, which define information that the employer wishes to procure to ensure that the design is developed in accordance with their needs and that they are able to operate the completed development effectively and efficiently. Suppliers will respond to the employer's information requirements with a BIM execution plan.

The minutes of the project Pre-start or Kick-off Meeting may also form part of the contract documents, although care must be taken to ensure that these minutes do not contradict anything contained in the other contract documents.

It should also be noted that the contractor's programme or baseline schedule is not included in the recommended set of contract documents. In many countries it is quite normal for this contractor's programme / schedule to be part of the contract documents, in which case the contractor is legally bound to maintain the programme / schedule. This can create its own set of difficulties in the construction stage:

a The programme / schedule is inextricably linked to the contractor's Method Statement, which may be confidential.
b If the employer wishes to issue a Variation Order, a new programme / schedule will have to be amended and approved by all parties. This would lead to very cumbersome contractual procedures.
c The contractor's responsibility is normally only to finish the project by the contract completion date, unless there are sectional completion dates. If the

programme / schedule is a contract document, then the dates for each individual operation become a contractual obligation.

Therefore, most standard forms of contract do not insist that the contractor's programme / baseline schedule is a contract document, but invariably insist that one is produced within a set period of time after commencement for approval or acceptance by the PM / engineer.

2.4 BIM: Building Information Modelling

2.4.1 BIM generally

BIM has been gaining considerable ground over the last few years and has become policy for many clients, especially government departments in the UK. It is essentially an approach to building design, construction and operation which is intended to change the way the industry uses information and communication technologies (ICT), which are now becoming much more powerful, accessible and user-friendly. The intention of BIM is that all project participants should have quick and easy availability and access to project design scope, schedule, and cost information that is accurate, reliable, integrated and fully coordinated.

The advantages of BIM are seen to be:

- increased speed of delivery of the project (i.e. time saved)
- better coordination of design (fewer errors and increased buildability)
- reduced costs of construction (money saved)
- increased productivity and efficiency of construction processes
- higher quality work (right first time)
- better client satisfaction (resulting in new revenue and business opportunities).

For each of the three major phases in the building life cycle – design, construction and operation – BIM offers all the project participants access to the following critical information:

- in the design phase – design, schedule, and budget information
- in the construction phase – scope of works, quality, schedule, and cost information
- in the operation phase – performance, utilisation, and cost-in-use financial information.

The ability to keep this information up to date and easily accessible to all who require it has always been a major hurdle in the acceptance of computerised techniques, since detailed data input has always been very time consuming and expensive. BIM addresses this issue by capturing the data when it is first input by the relevant design professionals, such as architects or engineers in terms of design data or cost consultants in terms of cost data in a fully integrated computerised format. Thus all project participants can be given a clear overall vision of the project, as well as the ability to make better quality decisions, thereby raising the overall quality, increasing the profitability, reducing errors and mistakes, and thereby increasing client satisfaction. BIM is therefore essentially a management approach rather than an off-the-shelf technology, although it does require suitable sophisticated technology for it to be implemented effectively and fully.

Therefore BIM is, essentially, the intersection of two critical ideas:

- keeping critical design information in digital form in order for it to be easier to update and share and of more value to the organisations both creating and using it;
- creating real-time, consistent relationships between digital design data and the information needed for efficient and effective operation of the building. This can save significant costs in the operation stage.

Unless BIM has been incorporated in the design phase of the project, there is little that can be done in the construction phase. However, if BIM is operational, there will be full concurrent information on building quality, schedule and cost available to the project managers on site. The cost consultant and contractor can accelerate the quantification of the building for estimating and value-engineering purposes and for the production of updated estimates and construction planning. The effects of proposed variations and value-engineering proposals can be studied and understood more easily, and the contractor can quickly prepare plans showing site use or phasing for the client, thereby communicating and minimising the impact of construction operations on the client's operations and personnel. BIM also means that less time and money will be spent on process and administration issues in construction because document quality is higher and construction planning better. The end result is that more of the owner's funds go into the building rather than into administrative and overhead costs and there will be less need for re-work or periods of delay as the coordination between trades and elements is fully integrated.

2.4.2 BIM benefits in the operations phase

In the operations phase of the building life cycle, i.e. when the construction has been completed and the building is 'working', BIM is designed to make available concurrent information on the use or performance of the building, its occupants and contents, the life of the building over time and the financial aspects of the building, such as rent, taxes and utilities costs. BIM is designed to provide a digital record of renovations and other physical changes to the building, which will help the sale or rental process by providing the documentation required for potential tenants. The systems may also be used to provide data for costs analyses of new buildings where the new buildings are of similar construction and use but in different locations. Physical information about the building, such as finishes, tenant or department assignments, furniture and equipment inventory, and financially important data about leasable areas and rental income or departmental cost allocations are all more easily managed and available. Consistent access to these types of information clearly improves both revenue and cost management in the operation of the building.

2.4.3 Future potential of BIM

When BIM is used effectively, designers will be able to make use of a project's digital design data to provide new services and therefore hopefully gain new sources of income.

Experienced clients are increasingly demanding digital models of their buildings, especially where they have a portfolio of buildings in their ownership. Designers and project managers may be able to offer new and expanded services, such as move management, energy analysis, digitally integrated cost estimating, and renovation phase planning, no doubt for additional fees.

In the future, BIM is expected to empower both design and construction professionals to work more collaboratively throughout the project delivery process, focusing their energy on more value added functions such as client requirements, creativity and problem solving, while computers do the tedious tasks of number crunching. Many experienced quantity surveyors will, however, have heard that one before.

But for real property owners and professional project managers, BIM holds great promise beyond merely improving productivity in the design and construction process. Ultimately, this approach and technology has the potential to enable the seamless transfer of knowledge from asset planning through design, construction, facilities management and operation, into the various disposal options. Whilst all parties involved in design and construction stand to gain from the adoption of BIM, it is the clients who will potentially benefit the most, through the use of the facilities model and its embedded knowledge throughout the economic life of the building.

This potential can only be realised if the information contained in the model remains accessible and usable across the variety of technology platforms likely over a

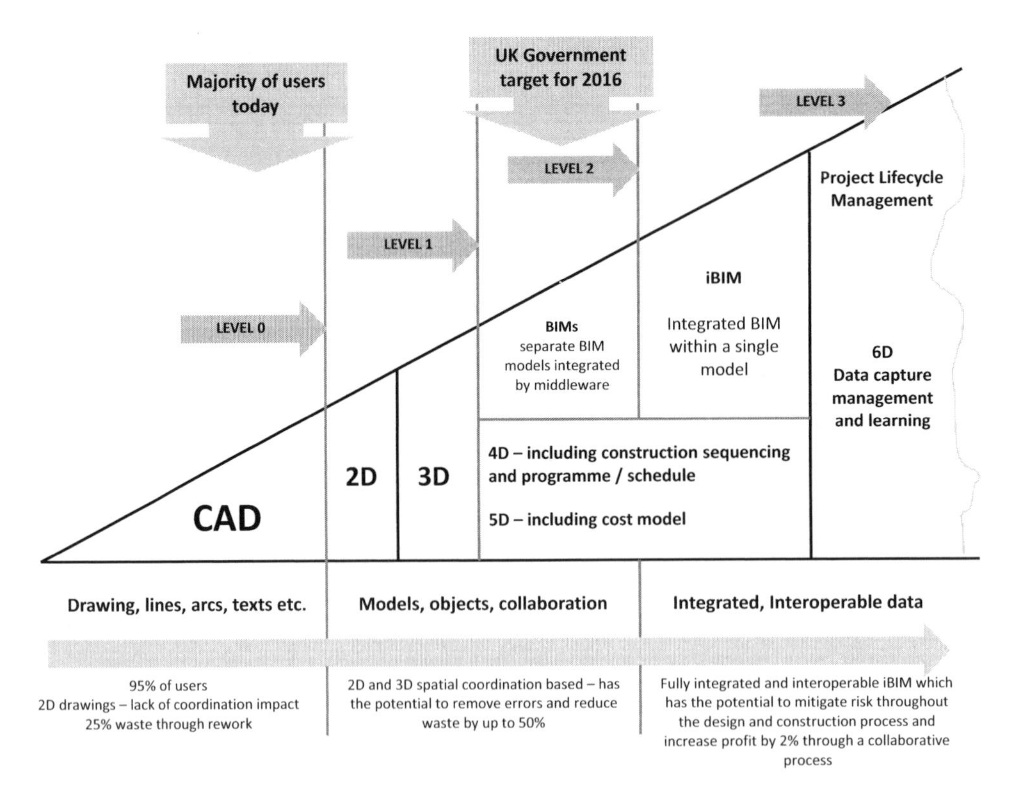

Figure 2.3 Development of Building Information Modelling (Developed from Bew and Richards 2008).

long period of time. Given the accelerating pace of technology, in 20 to 30 years our now state-of-the-art hardware and software applications will undoubtedly be outdated and obsolete. It is therefore essential that BIM is developed within a universal, open data standard to allow full and free transfer of data among the various applications as they are revised, updated and improved.

Figure 2.3 gives a generalised graphical view of the development of BIM through its various levels. As of today, virtually all projects are designed using some form of computer aided design (CAD) software package. These packages allow easy transfer and transmission of design information to the project parties who need it and can integrate with standard project collaboration software. In addition, most of the packages allow some form of 3D simulation of the design together with the ability to interrogate the detail of building components. The next stage is clearly to integrate the cost planning and scheduling / programming function into the package so that the three areas of time–cost–scope are covered in a single package. This is now where BIM has got to in 2015 / 2016. Future developments include operation and maintenance design, whole life costing, project life cycle management together with automatic feedback of data and information for both project learning and use on future projects.

2.5 Summary and tutorial questions

2.5.1 Summary

Communications are clearly of critical importance in effectively managing any organisation; and projects are no different. As projects have a defined start and end, with a particular scope of work to be delivered within that time frame, then effective communications are even more important in order not to lose any time in the Forming – Storming – Norming – Performing process (for a further discussion of this process, see any reference to Tuckman's Stages of Group Development).

Communications will either be verbal or written. Verbal communications clearly have the disadvantage of not being recorded, which is why many contractors use a system of CVIs – confirmation of verbal instructions – to convert the verbal communication into a written formal record. In both cases, the communication will be most effective when the giver and receiver have the same understanding of the message. This is also why many legal / contractual letters start with the phrase 'for the avoidance of doubt . . .'.

As we shall see in Part D of this book, a major imperative of construction contract administration and management is the keeping of records, for use as evidence or proof in any future variation or claim. As construction projects are both complex and fast moving, with letters, reports, minutes, instructions etc. being generated and communicated on a daily (even hourly) basis, therefore control of the communication and transmittal of these documents needs to be both organised and efficient. Document control has therefore become a project management industry in its own right.

The main controlling documents in a construction project are the 'contract documents', which set out the scope, time and cost of the project. It is therefore essential that these documents are clearly defined and understood by all parties. All construction management and supervision activity will refer to these documents as the baseline.

Furthermore, as the industry progresses and incorporates ever more complex information and communication technologies (ICT), the role of BIM will increase; and this can be seen with increasing frequency in the technical press. At present, BIM is relatively immature in its use in the industry, except for a relatively small number of mega projects. This is not surprising given the fragmented nature of the industry but things are changing slowly as the government puts more effort into this initiative. Internationally, the take-up is even slower and will struggle to make any headway in the tendering climate where lowest cost equals better value. Notwithstanding, BIM is clearly the future for project collaboration, communication and documentation.

2.5.2 Tutorial questions

1. Why is effective communication important in construction projects?
2. Discuss the main features of best practice in document control.
3. Give examples from the construction industry for each level of the hierarchy of documents.
4. Why is it important to have the contract documents agreed before commencement of construction operations?
5. Discuss the advantages and disadvantages of including the contractor's programme in the contract documents.
6. Outline the major principles of BIM.
7. BIM is just a glorified 3-D CAD. Discuss.
8. Effective communications and project control will never be achievable while the industry is so fragmented and lowest cost tendering is still the norm. Discuss the implications of this statement.

3 Risk identification and management

3.1 Introduction

A project risk can be any event (or indeed a series of events) that will have negative consequences on the project if it occurred. This negative consequence may be in terms of performance, functionality, time or cost.

The British Standard on Project Management (EN BS 6079-3:2000) defines risk as:

> An uncertainty inherent in plans and the possibility of something happening (i.e. a contingency) that can affect the prospect of achieving business or project goals.

Therefore the objective of risk management is to ensure the rapid identification of potential hazards (risks) within the project and to establish a clear process of assessment, control and mitigation of the identified hazards and risks. It is clearly important that any potential threats to the project should be eliminated or reduced to an acceptable level in order to allow the project to progress as smoothly as possible with minimum unnecessary interruptions. The underlying principle is that the major risks and their appropriate control measures are kept under regular review and reported to the project management team before risk management becomes crisis management. In other words, don't make a crisis out of a hazard.

3.1.1 Risks versus issues

Risks and issues are not quite the same thing, except that the exact nature of both is largely unknown before the project commences. With risks, there is usually a general idea in advance that there may be a cause for concern (because the hazard has been identified) and that something should or must be done about it. An issue tends to be less predictable, in that it can arise out of the blue with no warning and therefore be completely unpredictable.

3.2 Risk assessment and management

3.2.1 Risk assessment

It is clearly very important to identify hazards and potential risks before the project commences, and this is usually carried out in a risk management workshop; in order to prioritise these risks, there are several graphical tools available for use in this workshop. Firstly, a risk-impact matrix provides a useful framework which can then be used to proactively manage the risks with appropriate prearranged solutions. See figure 3.1 for an example of a risk-impact matrix with examples from an international project – note the size of the balloon represents the relative importance of the risk item. Secondly, a risk radar chart (figure 3.2) graphically illustrates which are the major risk factors, thereby allowing the risk management strategy to be developed. After all the potential risks have been identified using these techniques, they can be further assessed and prioritised using a risk-impact rating matrix (figure 3.3), although this gives broadly similar information to the matrix in figure 3.1.

Both of these charts can be developed within the project risk management workshop and should include all the project participants mentioned in Chapter 1. At this workshop, the strategy for dealing with the risk would be agreed and a delivery mechanism put into practice.

The issues in the top right hand corner of the matrix, i.e. those with high risk and high impact are clearly the priority to be managed and avoided. Similarly, those potential risks with either high risk or high impact will need a strategy for minimisation. Those risks in the bottom left hand corner could easily be dealt with if and when they occur as they would have a minimal impact on the project, if at all.

Figure 3.2 gives a slightly different graphical representation of the risks and would be better developed following the risk management workshop, as the separate risk issues would need to be established before the chart can be constructed. The number of 'spokes' would clearly increase as more risk issues are identified, so it is easy to see how this kind of chart could become quite complicated and cluttered. In larger schemes,

RISK \ IMPACT	**Insignificant** Minor problems easily handled by normal day to day processes	**Minor** Some disruption possible and financial cost	**Moderate** Significant time and resources required to deal with issue	**Major** Operations severely affected with major cost impact	**Catastrophic** Business survival is at risk
Almost Certain				Project not staffed in time	
Very likely					
Moderate chance		Poor communications		Language misunderstandings	
Unlikely during project					Major civil commotion / war
Very rare					

Figure 3.1 Risk-impact matrix.

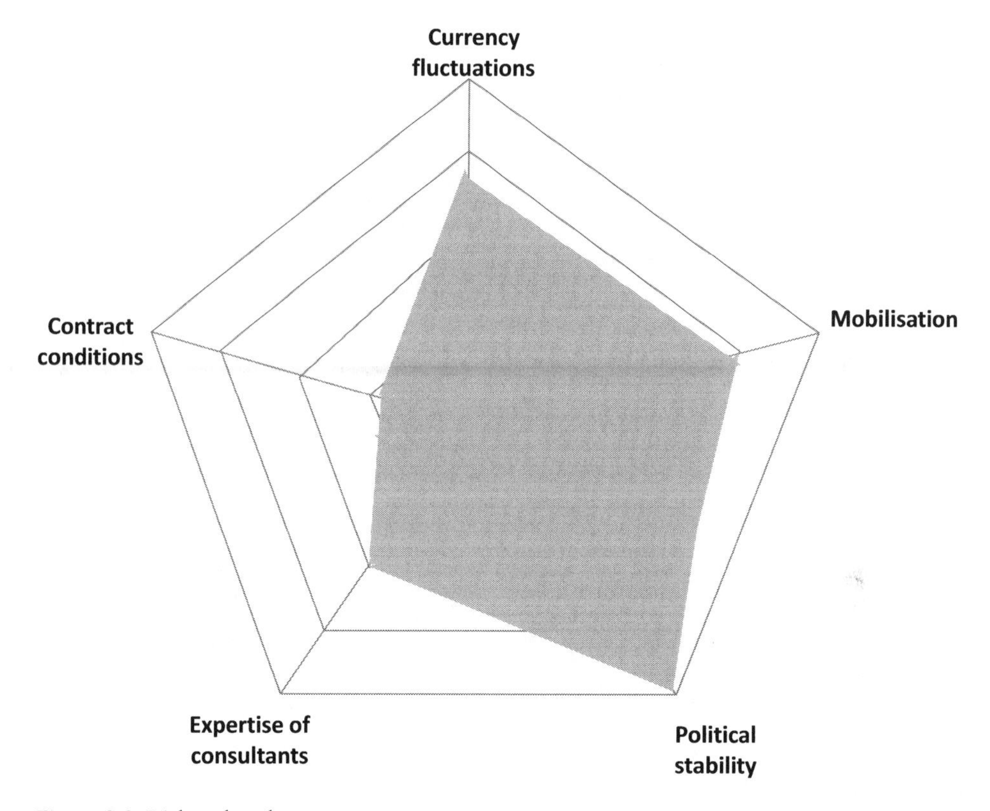

Figure 3.2 Risk radar chart.

separate charts could be developed for different sections / work packages of the project, or indeed for different contractors / sub-contractors employed to carry out the works.

The different risks associated with construction projects can be grouped into the various risk types shown in figure 3.4, which gives a general classification of risks within the four categories of technical risks, external risks, organisational risks and project management risks. There are many other classifications in different specialist textbooks and these may vary depending on the nature, location and characteristics of the project. Each client or individual project manager would be able to define the type of risk which will be common to their own circumstances and structure their risk breakdown system accordingly.

3.2.2 Risk response

When the risks have been assessed and prioritised, they clearly then need to be pro-actively managed. Figure 3.5 shows a standard Risk Assessment Plan where all the risks will have been identified and analysed in terms of their likelihood and impact. An appropriate risk response will be developed for each one and managed by the project team throughout the delivery stage of the project. This risk response will be written into the project's risk register which is a natural progression of the risk assessment plan but concentrating on the risk response and monitoring and control.

	Reputation of Delivery of Product or Service **D**	Health, Safety and Environment **S**	Commercial **C**
5 Threat to business survival and corporate credibility	Association with high profile or sensitive issues project may have critical impact on stakeholder interests. Not within normal business territory.	Multiple fatalities and/or major environmental incident involving public health / safety. Negligence and/or criminal liability including prosecutions.	Cost effect forces business liquidation.
4 Threat to future trading and core client /business objectives	Core stakeholder interests will not be protected by this impact on quality, programme or cost. Possibility of protestor action or media campaign. Major impact on goodwill.	Fatalities with criminal liabilities resulting in prosecutions. Environmental incident with protestor actions and adverse publicity. Legal and compensation costs.	Considerable cost effect with major impact on profits and share price.
3 Partial delivery of client/ business requirements	Impact on the project will result in only partial delivery of employer requirements. Media coverage may result in damaged stakeholder image	Major injuries to workers or third parties. Possibility of action for damages and/or nuisance.	High cost effect with impact on project and business profitability.
2 Late / inconsistent delivery of client / business requirements	Impact on the project will result in delayed delivery of employer requirements. Media coverage may result in strained stakeholder relationships.	Minor injuries to workers or third parties. Environmental impact requiring management response to recover.	Moderate cost effect with impact on project profitability.
1 Negligible impact	Negligible impact – can be managed to mitigate any effects.	Negligible impact – can be managed to mitigate any effects.	Minor cost effect absorbed in day to day finances.

Figure 3.3 Risk-impact rating matrix.

There are four major ways of dealing with the various risks that have been identified in a construction project:

1 *Avoid the risk* – by management action such as changing the method of work.
2 *Reduce the risk* – such as obtaining more information so the level of uncertainty is reduced or taking preventative or protective action. The use of PPE (Personal Protective Equipment) by workers on site is an example of this.
3 *Transfer the risk* – by taking out insurance or contractually transferring it to others via sub-contracts or separate work packages.
4 *Retain the risk* – take the chance yourself. If the hazard manifests itself – you pay.

3.3 Issue management

In the life cycle of any project, there will invariably be unexpected problems and issues that crop up from time to time. When these issues arise, the management team needs to be able to deal with them promptly so that they do not delay or seriously affect the successful outcome of the project.

Since most of these issues are, by their very nature, unexpected, then in order to deal with the issues quickly and effectively, an 'issue resolution process' should be put in place well before the project commences, in order to make sure that the project remains on schedule, on budget and meets its objectives.

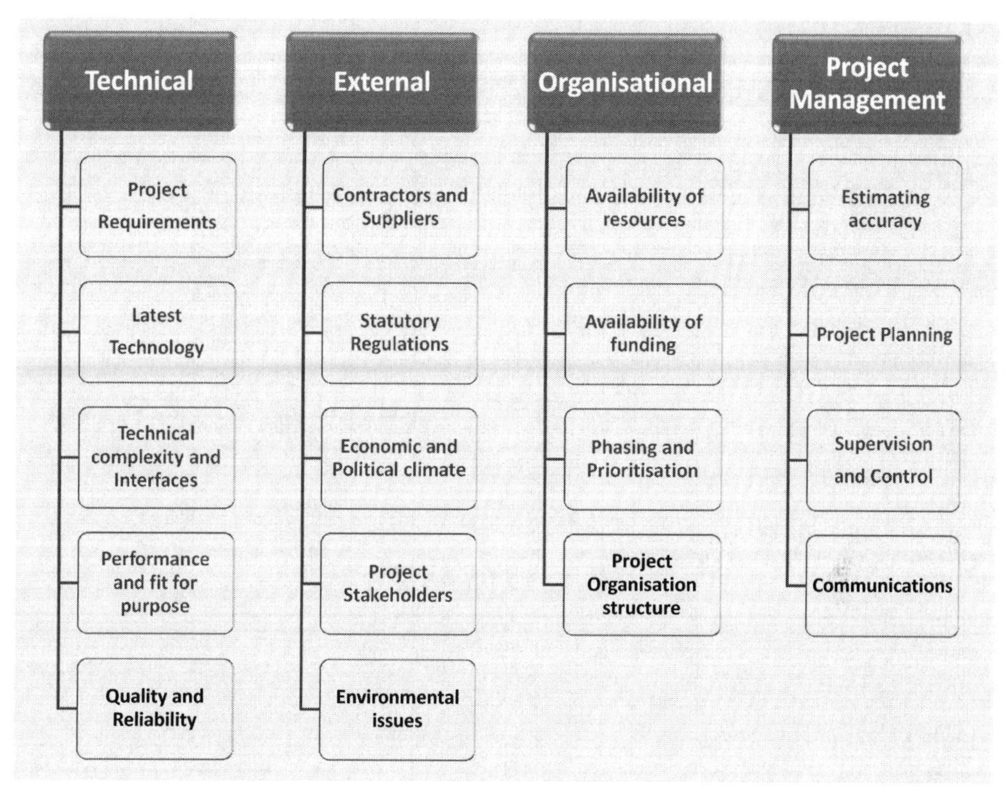

Figure 3.4 Risk breakdown structure.

Issue management is the process of identifying and resolving any issues which will crop up during the progress of the project. This can include problems with staff or suppliers, technical failures, material shortages etc. – all of which could have a negative impact on the project. If the issue remains unresolved, it will most likely create unnecessary conflicts, delays or even failure to produce the project deliverables.

An issue is therefore something that just happens. They are slightly different from risks as they may not have an effect on the cost or time element of the project, but nonetheless need to be proactively managed by the project team. For example, being unable to employ suitably qualified or skilled staff on a construction project is an identifiable risk. However, when one of your key staff is involved in a car accident, and hospitalised for three weeks, that becomes an issue. When it comes to issues, you have to deal with each of them as they happen. Issue management, therefore, is a planned process for dealing with an unexpected issue – whatever that issue may be – if and when one arises: in other words, planning to deal with the unplannable.

3.3.1 Issues management framework

The Issues Log should form part of an overall framework, or process, for dealing with those issues and will help the project team understand what to do with issues once they have been identified and logged. The advantages of having an Issues framework include:

PROJECT RISK ASSESSMENT PLAN																
	RISK IDENTIFICATION								QUANTITATIVE ANALYSIS			RISK OWNER	RISK - RESPONSE		MONITORING & CONTROL	
	Status	ID #	Date identified in project phase	Risk Event	SMART description	Risk Trigger	Impact area	Affected WBS	Probability	Impact	Risk Matrix		Strategy	Action to be taken	Status interval or milestone	Date, Status & Review comments
(A)	(B)	(C)	(D)	(E)	(F)	(G)	(H)	(I)	(J)	(K)	(L)	(M)	(N)	(O)	(P)	(Q)

Figure 3.5 Project risk assessment plan / risk register.

- how to assign responsibility for resolving issues that occur and to whom;
- when to escalate issues to a higher level of management. This can be done by creating a matrix of potential business impact versus issue complexity to help decide which issues could and should be escalated;
- which criteria will determine an issue's priority status?
- who will set the target resolution date?
- how will issues be communicated within the team and externally? For example, regular meetings, log checks, status update emails, press releases, public statements etc.?
- how can issues be disaggregated if several occur simultaneously on a project?
- if change orders are needed, how will they be handled?
- when the resolution affects the budget or schedule, how will the overall cost and time effect be updated and who will be responsible?
- will the issue generate a contractor's claim? If so, the claims manager should be involved at an early stage.

One of the key challenges of issues management is to resolve the problem quickly and then move on, creating as little impact to the project as possible. An issues framework provides a structure for making decisions as and when they arise. Remember also that nothing is done in isolation, so be aware of the effect on other project procedures – cause and effect analysis.

It is also important to ensure that all issues that have arisen are covered in both the Progress Report (see section 4.4 in Chapter 4) and the Completion Report (see section 4.7 in Chapter 4), which is where the lessons learned can be captured to help with future projects. The more you learn about issues, the better prepared you will be for the next project. Clearly, some issues may occur again and again, so by recording what has been learned from previous projects, subsequent project teams may find it easier to identify likely issues, and be able to resolve them successfully. Other issues might be part of a risk pattern that can be identified and managed through the risk management framework.

An issues management process therefore allows a robust method of identifying and documenting issues and problems that occur during a project. The process also makes it easier to evaluate these issues, assess their impact and decide on a plan for resolving them. An Issues Log is designed to capture the details of each issue, so that the project team can quickly review the status and allocate somebody to be responsible for resolving it. Adding an issues management framework will give a comprehensive plan to deal with issues quickly and effectively, which can be used to refine and improve future projects.

3.3.2 Issues Log

Issues – i.e. problems, gaps, inconsistencies, or conflicts – need to be recorded when they happen. When an Issues Log is created, this becomes a tool for reporting and communicating what is happening on the project. The log also has the effect of making sure that issues are indeed raised, and then investigated and hopefully resolved quickly and effectively. Without a defined process, there is a risk that issues will be ignored or not taken seriously enough – until it is possibly too late to deal with them successfully. An Issues Log allows the following:

- having a safe and reliable method for the team to raise issues;
- tracking and assigning responsibility to specific people for each issue;
- analysing and prioritising issues more easily;
- recording all issue resolutions for future reference and project learning;
- monitoring the overall project health and status.

Issues Logs can be created manually by building a spreadsheet or database from commercially available shell software packages, or by purchasing issue management software from a wide variety of specialist vendors. However, the success of an issue management process does not necessarily depend on which tracking mechanism is used, but rather on the type of information tracked. The following information should be included in an Issues Log.

Issue type

The log should define the categories of issues that are likely to be encountered on the project, which will help in tracking the issues and assigning the right people to resolve them. The descriptions should be reasonably broad, such as:

- *Technical* – i.e. issues relating to a technological problem in the project.
- *Business process* – i.e. relating to the project's organisational design and operation.
- *Change management* – i.e. relating to business, customer, or environmental changes.
- *Resource* – i.e. relating to plant and equipment, materials, or personnel problems.
- *Third party* – i.e. relating to issues with vendors, suppliers, or parties outside the supply chain.

In addition, the identifier of the issue should be recorded (possibly for linking with thank-you awards if the issue is not part of their normal duties) as well as the timing that the issue was identified, to assess whether there is sufficient time for a proper resolution. The description of the issue should provide details about what happened as well as the potential impact. If the issue was to remain unresolved, the parts of the project that will be affected should be clearly identified and the responsible people notified.

Priority

A priority rating should be assigned to the issue, in the same way as for risk analysis. For example:

- *High priority* – critical issues which will have a high impact on project success, and have the potential to stop the project completely.
- *Medium priority* – issues which will have a noticeable impact, but won't stop the project from proceeding.
- *Low priority* – issues which will not affect activities on the critical path, and probably will not have much impact as long as they are resolved at some point.

Assignment / owner

The person responsible for resolving the issue should be determined in the log. This person may or may not actually implement a solution, but they will be responsible for tracking it and ensuring that it is dealt with according to its priority.

Target resolution date

Establish the deadline for resolving the issue.

Status

The progress of the resolution should be able to be tracked with a clear label identifying the issue's overall status. For example:

- *Open issues* – the issue has been identified, but no action has yet been taken.
- *Investigating* – the issue, and possible solutions, are being investigated.
- *Implementing* – the issue resolution is in process.
- *Escalated* – the issue has been raised to management or the project sponsor / steering committee, and directions or approval of a solution is pending.
- *Resolved* – the resolution has been implemented, and the issue is closed.

3.4 Dealing with risk – insurance, bonds and guarantees

3.4.1 Insurance

Most standard forms of construction contract typically prescribe that project insurances will be obtained by the contractor and the costs are accounted for in the preliminaries section of the pricing document (e.g. Bill of Quantities). In some cases, the employer will provide the project insurance themselves; for example, if the employer is an insurance company or financial institution itself. In virtually all cases, the following require to be insured:

a The works itself (both permanent and any temporary work)
b Contractor's plant and equipment
c Third parties including injury to persons and damage to adjacent property
d Accident or injury to workmen and contractor's personnel
e Damage to employer's property.

The relevant provisions will clearly depend on the nature of the project, including location, size, specific risks etc., but most construction projects will carry the following insurance policies:

- *Contractor's all-risk policy (CAR)* which usually covers any loss or damage to the works, materials on site, contractor's plant and equipment, temporary site accommodation, personal property (e.g. tools owned by tradesmen). The CAR policy is normally taken out in the joint names of the contractor and employer and may also include sub-contractors.
- *Public liability policy* which insures the contractor against the legal liability created by death, injury or damage caused by the activities on site.
- *Employer's liability policy* which covers staff and labour for the head office and each separate site.
- *Professional indemnity (PI) insurance.* This is the insurance that will be required when the contractor has any design liability and can also cover liability for negligence in any supervision duties.

Prior to any work commencing on site, the contractor will be required to show evidence that all insurances are in place and to provide copies of the policies to the contract administrator, employer or PM / engineer. The employer sometimes insists on approving the terms and conditions of the insurance policy but this is becoming less common, especially where the insurance is taken out with a reputable and substantial insurance company and also it is implicit that the insurances shall comply with the contract requirements, and of course the employer is entitled to insist on the contractor's compliance. The employer should formally notify the PM / engineer whether or not the contractor's insurance submissions are compliant. The FIDIC Conditions of Contract state:

- in the event of compliance: the Engineer can allow the Works to proceed and certify any payment entitlements under the Contract for insurance;
- in the event of non-compliance: the Engineer can issue to the Contractor notice of requirement to remedy this default within the time stipulated in the Contract or, if none is stated, within a reasonable time.

If, under the contract, or under a Pre-construction Services Agreement (PCSA), the contractor designs major parts of the works, then as stated above they may also be required to take out professional indemnity (PI) insurance against any design liability in the same way as independent architects or design engineers are required to do.

If the contractor fails to comply with the obligation to take out project insurance or does not provide the appropriate evidence, there is normally a provision that the employer may take out the insurances themselves and recover the premiums from the contractor by deductions from certified payments. If no insurance is taken out at all, but the contractor is contractually required to do so, then the contractor will be liable for any costs which follow an insurable event which should have been recoverable under the insurances that the contractor failed to take out. It is normally the responsibility of the PM / engineer to ensure that the relevant insurances are in place before giving the contractor possession of site and the Notice to Proceed which will all be covered in the agenda for the Pre-start / Kick-off Meeting (see section 4.3 in Chapter 4).

Although it is the employer's responsibility to accept (or not) the contractor's insurance submissions, the employer will usually request the PM / engineer's assistance and advice.

Project insurance is a complicated area and clearly there can be a major financial effect if the process is not carried out properly and the insurance company refuses a claim due to a technicality. Therefore, the PM / engineer and contract administrator must ensure that:

- Satisfactory evidence of insurances is in an appropriate format, such as valid certificates from the insurer.
- Insurance premiums have been paid and the required insurances become fully effective by the date that the evidence is required to be submitted.
- The insurance policies must be scrutinised in detail against the stated requirements of the contract in all respects including:
  - Limits of insurance per event and in aggregate (except where the contract does not permit limitation of the number of events);

- Limits of deductibles;
- Limits of location: locations to be included would be the contractor's site office installations, maintenance facilities, storage and lay-down areas and fabrication / manufacturing sites (which might be in a different country), source locations (e.g. quarries), access roads and goods in transit. The 'site' as described in the contract particulars may not be sufficient for insurance purposes;
- All sub-contractors are covered by the contractor's insurances. Insurance amendments may be required whenever new sub-contractors are employed, and the employer with the PM / engineer need to monitor that the contractor provides such amendments during the course of the contract. If sub-contractors are not covered by the contractor's insurance policies, then the contractor would be required to demonstrate that the sub-contractors have their own effective insurance complying with the main contract requirements;
- Policy exclusions should not be accepted, other than those identified in the contract as 'Employer's Risks', unless the contractor demonstrates that insurance coverage of such other excluded risks is not available at commercially reasonable terms;
- The terms of duration of validity of the insurance policies: the contractor is required to maintain all insurances throughout the execution of the works and, to a lesser extent, during the defects correction period, but as insurances are often taken out on an annual basis, then for contracts over 12 months duration the insurances will need to be renewed. The insurances will also need to be extended for all extensions of time awarded by the PM / engineer.

Therefore, as the contract progresses, the employer, together with the PM / engineer must monitor that the contractor maintains adequate insurance in accordance with requirements of the contract.

3.4.2 Performance bond

A performance bond is a form of financial guarantee that the contractor will perform the works that they have contracted to perform. The bond is normally valued at 10 per cent of the contract sum and will be issued by a recognised financial institution (termed the *Surety*). The amount of the bond will be payable to the employer in the event of the contractor defaulting on the performance of their obligations under the contract by failing to complete the works in accordance with the terms and conditions of the contract. The performance bond may be an *on-demand bond*, in which case the employer may call in the bond on demand and there is no requirement to prove default, or it may be a *conditional bond*, in which case the amount will only be paid by the surety when the conditions have been proved (i.e. non-performance). Many international clients insist on the bond being on-demand as the proving of a default can take a long time, which the contractor will naturally contest. On-demand bonds are considerably more expensive than conditional bonds which will be paid by the employer anyway via the project preliminaries. Even though the employer will be paid on demand, they would be well advised to think very carefully about calling in such bonds as it may have an effect on their credit rating with the financial services industry as well as on the willingness of established contractors to work with them in the future.

3.4.3 Advanced payment bond

Most modern standard forms of construction contract include a provision to pay the contractor a percentage of the contract sum before they have started any work, which is intended to assist with their mobilisation and set-up costs. This is most common in international contracts, where the contractor may be mobilising from another country and will need to import plant, equipment and labour as well as constructing accommodation and welfare facilities for their staff, batching plants for the production of concrete, storage and lay-down areas etc.

The advance payment is normally 15 per cent of the contract amount and paid within a specified time period after the contract comes into force and before physical commencement of the works. The advance payment is normally repaid by a percentage reduction in interim payments throughout the project, until the full amount of the advance has been repaid. The contractor will also normally be required to provide an advance payment bond, which is a guarantee bond, purchased from a bank or other financial institution whereby the client will be paid back the advance payment by the guarantor (bank) in the event of the contractor's default or their contract being terminated. In order to lower the costs of the bond, they may be taken out on a reducing balance basis, so that as the advance payment is paid back via the Interim Certificates, the value of the bond can be reduced accordingly. When the advance payment is fully paid back, the bond value will be zero and therefore will expire – see figure 3.6. See also section 5.4.1 for additional discussions on advance payment bonds.

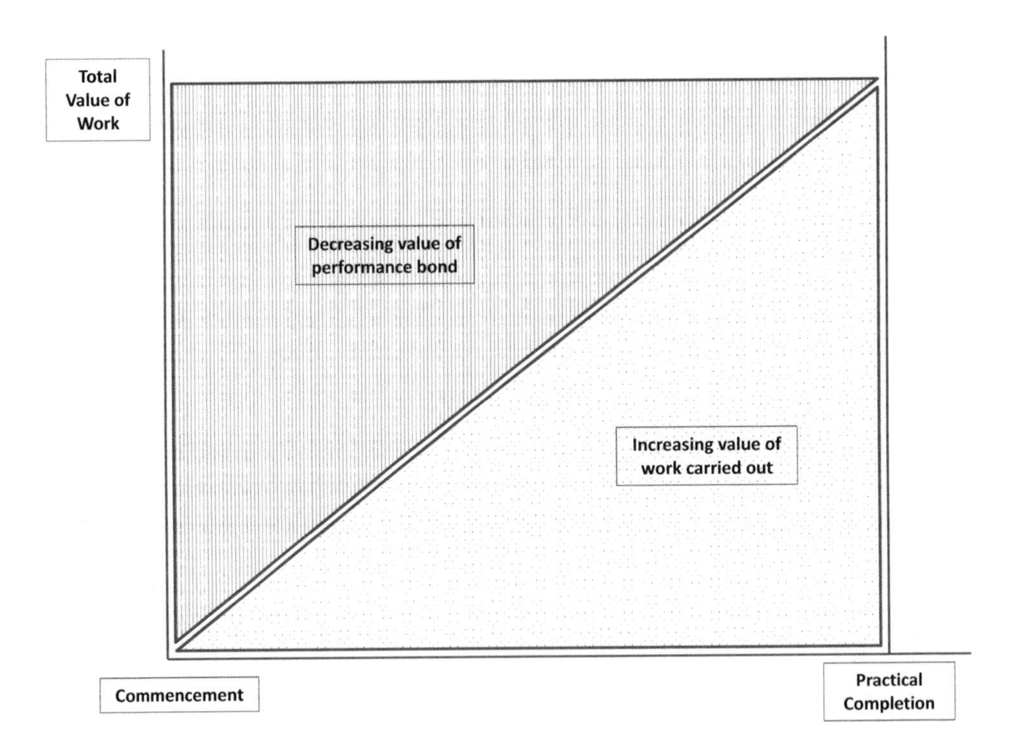

Figure 3.6 Reducing value of advance payment / performance bond throughout the project.

3.4.4 Retention bond

This is a much less common type of surety which again is obtained from a financial institution and replaces the need for the employer to retain monies which have been properly earned by the contractor. Clearly, this will help the contractor's cash flow during the works as any interim payments will not be subject to deductions. Upon any default by the contractor, the employer may call in the bond (which could again be on-demand or conditional), which would be paid to the value of an agreed proportion of the contractor's gross valuation at the date of the default. Contractors would much prefer to provide a retention bond to the employer, as this is a far cheaper alternative than having up to 5 per cent of their gross valuations held back by the employer (as the percentage profit on the majority of construction contracts is less than 5 per cent anyway, this could mean the contractor is not making any profit whatsoever until the repayment of the retention fund).

3.4.5 Payment bond

This is an even less common type of surety as it is required to be taken out by the employer for the benefit of the contractor. The purpose is to guarantee the valid interim payments and may be insisted on by the contractor for an employer who may have a history of late or poor payments. A more common solution to this issue is a *parent company guarantee*, if the employer is part of a holding company structure or is owned by a more substantial organisation. This can be particularly effective when the employer for a particular project is a shell company which has been created for the specific purpose of construction of that particular development. In these circumstances, the employer may only have the minimum legal requirement of finance – which could be as low as $100 – but may be signing a multi-million dollar contract for construction. Clearly, the contractor will wish to have some assurance that all due payments will be paid.

3.5 Summary and tutorial questions

3.5.1 Summary

All events in the future are uncertain. There is therefore a risk that what was planned to happen may not happen. During the design and tender stages of a project, the design of a new building is developed, together with a plan / schedule of how it will be constructed, together with the anticipated costs. There are plenty of things that can happen to disrupt this plan and the process of risk identification and management is the way in which some degree of mitigation can be exercised over the uncertainty. The normal process of risk management covers three stages:

- Risk identification
- Risk assessment / analysis / evaluation
- Risk mitigation / response.

Techniques used in each of these stages are covered in this chapter, but this can only be achieved for risks (or hazards) that are known or knowable before the event.

Factors which cannot be identified, assessed or mitigated before the event are termed 'issues' (section 3.3). Issue management requires a slightly different set of techniques, which can be just as crucial but less defined.

That's all very well, but the subject of risks and issues must be dealt with somehow. The construction industry has therefore developed certain products as part of the risk response – insurances, bonds and guarantees, which are normally provided as:

- Project insurance policies
- Performance bond
- Advance payment bond
- Retention bond.

3.5.2 Tutorial questions

1 What is the difference between a 'risk' and an 'issue'?
2 Discuss the major techniques used to assess risks in a construction project.
3 Give examples for each box in the risk-Impact matrix shown in figure 3.3.
4 Complete four risk issues in the risk register in figure 3.5, commenting on the effectiveness of the chosen monitoring and control mechanisms in columns (P) and (Q).
5 Comment on the major differences between the types of insurance policies covered in section 3.4.1.
6 Discuss the difference between an on-demand bond and a conditional bond.
7 Why would an employer require an advance payment bond?
8 Many international employers require a retention bond and also maintain a retention fund. Comment on the cash flow implications of this requirement.

Part B

Managing the construction stage

Performance and relationships

Part A covered the beginning of the construction stage of a project, i.e. setting up, contractor mobilisation and kick-off. When the contractor has commenced the execution of the scope of works, this execution needs to be managed to ensure it is carried out on time, within budget and to the required quality – the triangle in section 1.1.

This stage is therefore about delivery. Effective contract management during this stage will cover issues of performance, payment, variations and changes, and subcontracting. In addition, as major projects can often take a long time to both design and construct, design decisions may become outdated and new products may appear on the market which offer cheaper and/or better solutions to the employer's requirements. Therefore continuous value analysis or value engineering should also take place during this stage.

4 Supervising the contractor's performance

4.1 Introduction

This is possibly the most important aspect in effective contract management / contract administration. The contractor clearly needs to perform efficiently and effectively for the project to be successful for both themselves and the employer and management control on this performance can be viewed as a circular activity or a 'loop' as seen in figure 4.1. At the commencement of the construction stage there is the original plan, set out in the contract documents, which has been agreed by the employer as what they want to be built, by when and for how much. All subsequent activity in the project should be directed at achieving these requirements. The project team and especially the contractor 'performs' the works which is supervised and controlled by the PM / engineer together with the project management and supervision teams. This is all fairly straightforward and obvious, and what this chapter covers is the tools that have been developed to control the contractor's performance and ensure it achieves the original requirements of the employer as amended by subsequent variations and changes.

The PM / engineer's staff responsible for the supervision of the contractor will include technical engineers and construction managers to oversee the progress of the work on a daily basis. As far as this book is concerned, i.e. contract administration, the main tools for control are regular progress reports and regular progress meetings. The 'Control' box in figure 4.1 gives a list of these reports and meetings which should occur.

The various reports required in order to monitor and control the progress of the project should be set out in the Pre-start / Kick-off Meeting and will include:

- Inception Report (by the PM to the employer)
- Pre-start / Kick-off Meeting
- Project Progress Reports (usually monthly and / or quarterly)
- Project Progress Meetings
- Updating the project programme / schedule
- Construction Completion Report(s)
- Performance Report(s) on Contractor(s)
- Service Contract Completion Report.

4.2 Inception Report

An Inception Report is typically required to be produced by the project manager as part of their contract with the client and normally submitted within three months of

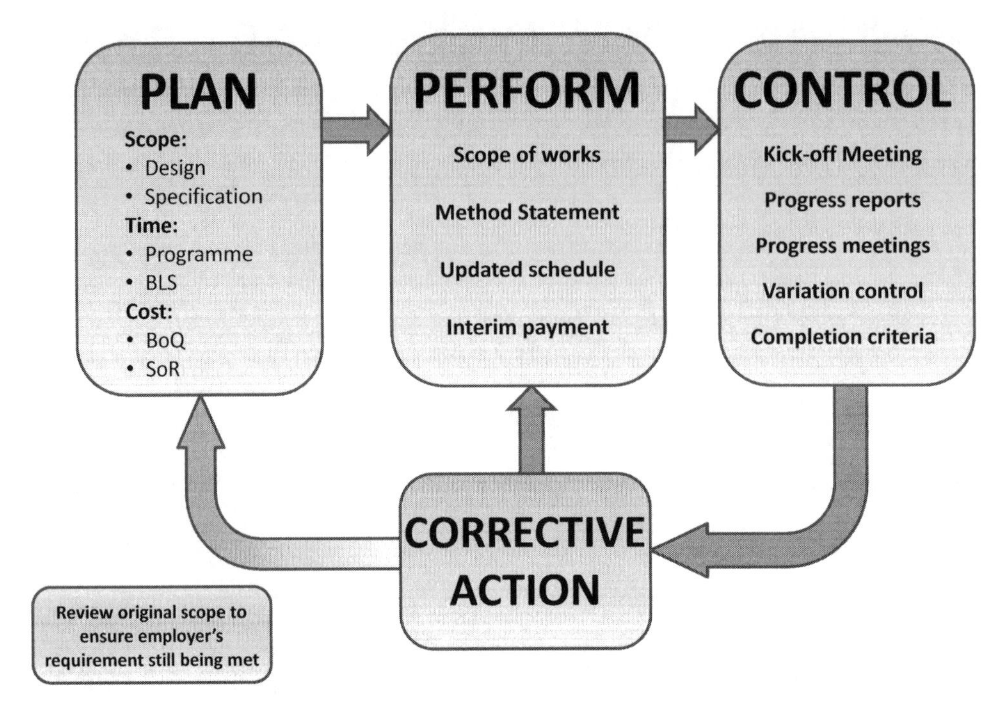

Figure 4.1 Construction project control loop.

commencing the project management services. The format and detail of the Inception Report will depend on the requirements of the particular client and an example / template is given in table 4.1. The main purpose of the Inception Report is to demonstrate the project manager's understanding of the project, by describing the start-up actions and plans to fulfil the client's project objectives. In general the Inception Report should:

- summarise the project background, objectives and status;
- summarise the background and purpose of the PM's Service Contract;
- provide a proposed project organisation showing reporting lines and interface between the various parties, i.e. Funding Institution(s), Client, Consultants, Subconsultants, site offices and contractors;
- describe the PM's mobilisations of project staff, facilities and equipment (to date and projected to anticipated demobilisation);
- describe the PM's activities (to date and projected) until the end of the project;
- summarise the status of preparations for or execution of the Works Contract(s);
- provide a Project schedule for the Works Contract(s) and the PM's activities.

An Inception Report also provides the opportunity to identify:

- changes in project objectives, scope or plans since the PM's original quotation / proposal;
- anticipated problems which could affect PM's scope and/or schedule of services.

Even though an Inception Report may not be formally required by the employer, it is still good practice to prepare one for internal use or possibly as the first Project Progress Report – i.e. giving the baseline situation. An example of a table of contents for an Inception Report is given in table 4.1.

4.3 Pre-start / Kick-off Meeting

The Kick-off Meeting is the first official meeting of the project team who will be working together on the realisation of the project. The agenda usually covers introductions, statement(s) of mission, and how the project will be organised, in terms of teams or working-groups and the timetable of subsequent project meetings will be set here. The meeting is also called a Start-up Meeting or Pre-start Meeting, or Pre-commencement Meeting depending on the country of the project. The meeting should take place just after issuing the Notice of Commencement Date so that all parties are aware that the clock has started ticking. It is essential that formal minutes are taken at this meeting, which are often referred to later in the project if (or when) there are any disputes or disagreements between the parties. See table 4.2 for a standard agenda of a Kick-off Meeting.

4.4 Project Progress Reports

At the start of the project the PM / engineer should prepare the template for the Project Progress Report (see table 4.2) and, as the project leader, will have full responsibility for the reports as they are produced. On large, multi-contract projects where there may be resident engineers for each Works Contract, the project manager should require each resident engineer to prepare a draft of the sections of the report relating to their particular section, following the established standard report format. The project manager will, however, still retain editorial privilege to amend and mould the draft sections into a consistent and coherent report for the overall project.

However, although the PM should have ownership of the Project Progress reports, the information and data will come from the contractor(s) and the majority of the standard forms of contract require the contractor to submit periodic (usually monthly but also weekly) progress reports. For example, FIDIC now requires the contractor to submit monthly progress reports within 7 days after the last day of the period to which each report relates. The detailed requirements are quite thorough and can be summarised as:

- charts and description of progress, including design, contractor's documents, procurement, manufacture, deliveries to site, construction, erection and testing, and works of each subcontractor;
- photographs showing the status of manufacture and progress on site;
- for each main item of manufacture: the name and location of manufacturer; percentage progress; actual or estimated dates of commencement of manufacture; contractor's inspections, tests, shipment and delivery to site;
- records of contractor's personnel, plant and equipment;
- copies of quality assurance documents, tests results and certificates for materials;
- list of notices of any claims;
- safety statistics; including details of all hazardous incidents;
- comparisons of actual and planned progress; including details of events or circumstances which may jeopardise completion in accordance with the contract, including any measures to overcome the potential delays.

Table 4.1 Contents of a Project Inception Report

PROJECT AT [PROJECT HOME] **FOR** **[EMPLOYER NAME]**
INCEPTION REPORT

1 PROJECT BACKGROUND

Purpose, scope and objective of the project, details of employer and any funding institutions.

2 PRECONSTRUCTIONS ACTIVITIES ALREADY CARRIED OUT

Design, statutory approvals and permits, land acquisition, praparotory site works, preparation of tender Documents, prequalification of contractors, preparation of Tender Documents etc.

3 BACKGROUND OF CONTRACT FOR PM / SUPERVISION

Prequalification,Requests for and submission of proposals selection.

4 CHANGES TO PM SCOPE OF WORKS / TERMS OF REFERENCE AND PROPOSED STAFFING

Necessary changes to SOW / TOR arising form changed project circumstances since proposal submission and / or signing of PM services contract and proposed changes to project staffing and schedule of services (if any).

5 PROJECT ORGANISATION

Description of how project will be organised, preferably with organisation chart in Appendix.

6 PROJECT SCHEDULE

Gantt chart of main project / Contract phases with identification of critical activities and deliverables.

7 CONSULTANT'S MOBILIZATION

Current and planned mobilisations as Gantt chart.

8 CONSULTANT'S PROGRAMME AND ACTIVITIES

Descripition of Consultant's required project activities, for example:

- *8.1 Prequalification of contractors*
- *8.2 Format of Tender documents*
- *8.3 Scope Works Contract No. 1*
- *8.4 Tender documents for Works Contract No. 1*
- *8.5 Scope Works Contract No. 1*
- *8.6 Tender documents for Works Contract No. 2*
- *8.7 Environmental considerations*
- *8.8 Land acquisition and Utility relocations*
- *8.9 Materials source surveys*
- *8.10 Economic Evaluation*

APPENDICES

A PROJECT ORGANISATION

B PROJECT SCHEDULE

C CONTRACT MOBILISATION

Table 4.2 Agenda items for a Project Kick-off Meeting

PROJECT AT [PROJECT NAME] FOR [EMPLOYER NAME]	
AGENDA ITEMS FOR KICK-OFF MEETING	
1	The role and authority of each entity participating in the Contract.
2	Appointment of the 'Engineer' and or 'Engineer's Representative' (as appropriate).
3	Appointment of the 'Contractor's Representative'.
4	Where the Contract provides for delegating duties and authorities to persons, these should be clearly established.
5	Status of availability for 'Access to and Possession of site'.
6	Requirement and procedures for obtaining 'Contruction License'.
7	Status of availability for 'Issuing Drawings'.
8	Status of contractor's 'Performance Security' and 'Advance Payment Security' .
9	Status of contractor's 'Insurance'.
10	Agreed 'Commencement Date'.
11	Requirements for 'Safety, security and Protection of the Environment'.
12	Requirements for 'Quality assurance and control'.
13	Status of the 'Works Programme', Key dates for information and submissions, periods for approval, long delivery periods and special problems.
14	Requirements for consent and approvals of any subcontractors.
15	Works or materials to be provided by the Employer.
16	Procedures for measurement, notices, instructions, submissions and responses.
17	Procedures for interim valuations, certifications and payments.
18	Procedures for monitoring the progrss of the works.

Previous versions of many of the standard forms of contract gave the PM / engineer little power to require the contractor to submit records of personnel, plant and equipment and / or monthly reports. However, later versions do provide these powers and some (e.g. FIDIC) allow the engineer to withhold an Interim Payment Certificate until the contractor has submitted the appropriate progress report.

Therefore, the PM / engineer should, at the beginning of the contract, remind the contractor that:

- Monthly progress reports are to be submitted for each calendar month within 7 days after the last day of the month to which the reports relate.
- The first monthly progress report is to be submitted within 7 days after the last day of the first full calendar month following the commencement date. The period to which this first report relates shall be the period commencing from the commencement date. (For example: if the commencement date is 15 April, the first report, covering the period 15 April to 30 June, is required to be submitted by 7 July.)

- Interim Payment Certificates in respect of the Contractor's Statements will be withheld until the contractor has submitted a monthly progress report for the period to which a Contractor's Statement relates.

Should the PM / engineer disagree with any facts of the contractor's monthly reports, this must be promptly recorded in writing to the contractor, copied to the employer, promptly notify such disagreement in writing to the contractor in an attempt to obtain the contractor's agreement to submit corrected records.

In the event that the contractor fails to justify his submitted records and disagrees to amend them, confirm the engineer's disagreement and the engineer's records to the contractor (copied to the employer).

The term Project Progress Report does seem a bit dated in this fast moving internet age, so the analogous concepts of 'dashboards' and 'scorecards' appear to be more *de rigueur* and required by current fashion and practice. An advantage of dashboards and scorecards is that the major items of performance can all be shown together in one view, which makes remedial action easier to put into effect – much like a car dashboard is to the driver.

As included in the Appendices for the monthly progress reports, *Progress Photographs* should convey an overall impression of works progress. Pictures looking down manholes may be technically interesting to some but do not convey an adequate overall picture of the project progress.

The main purpose of the reports is to record progress, not defects: although pictures of defects might sometimes be appropriate to support narrative within the reports dealing with such issues. Pictures should be taken from the same vantage points each month to illustrate progress. The contract documents may also require the contractor to submit monthly progress photographs, from which a selection of the best quality may be chosen for inclusion in the regular progress reports.

The reports may also include minutes of meetings and any selected correspondence during the reporting period. As the main purpose of progress reports is to inform the client of day-to-day progress and to keep a record in case of future claims, the inclusion of minutes and significant correspondence will ensure it is a more comprehensive insight.

Architect's duty to inspect

Many clients often have the expectation that the architect (or engineer) will supervise and inspect the work to ensure that the contractor carries out their duties properly and may also expect the architect to be responsible if the contractor fails to perform properly. The exact nature of the inspection duties clearly depends upon the terms of the professional services contract between the architect / engineer and the client, but will invariably include an inspection clause such as:

> Architect (Engineer) shall visit the site as required in order to monitor the progress and quality of the portion of the Work completed and to determine if the Work is being performed in accordance with the Contract Documents.

Therefore, if the contract requires the architect or engineer to periodically visit the work site and inspect the works, this inspection must be performed in a professional manner. If a defect is not noticed that would have been discovered through a reasonably diligent

Table 4.3 Contents of a Project Progress Report

PROJECT AT [PROJECT HOME] FOR [EMPLOYER NAME]	
MONTHLY PROJECT PROGRSS REPORT	
SECTIONS	
1	Project Data sheets (Names of parties, Commencement Date, Expected Completion Date, Project Budget etc.)
2	Brief description of the project (multilingual if required on international projects)
3	Progress of the works (descriptive and in numbers)
4	Delays, claims and contractual issues (raised, pending approved, rejected, resolved)
5	Submittals (Shop drawings, materials samples etc.) and quality issues
6	Variations and Change Orders (submitted, reviewed, approved, rejected)
7	Invoices and payments (Submitted, reviewed, approved, released)
8	HSE issues
9	Any other outstanding issues
10	Project risks (raised, assessed, allocated, resolved)
11	Comments and summary by PM
12	Conclusions
APPENDICES	
A	Progress photographs / multimedia files
B	Layout plans and updated project programme / schedule
C	Contractor's staff, labour and equipment
D	Material and equipment approval log
E	Shop drawing approval log (if appropriate)
F	Payments log
G	Claims log
H	List of approved sub-contractors
I	List of correspodences

inspection, the architect may be deemed responsible and consequently held liable in the tort of negligence.

Another issue that arises in this context is whether the architect can be held liable for safety violations. In other words, does the architect who is engaged to provide ordinary site services have any responsibility for injuries to people or property when the primary cause of the harm is contractor or sub-contractor negligence? The unsatisfactory answer to that question is maybe, as the court decisions on design professionals' liability arising from contractor errors are inconsistent at best and the outcomes can be very case-specific which makes it difficult to predict results. Still,

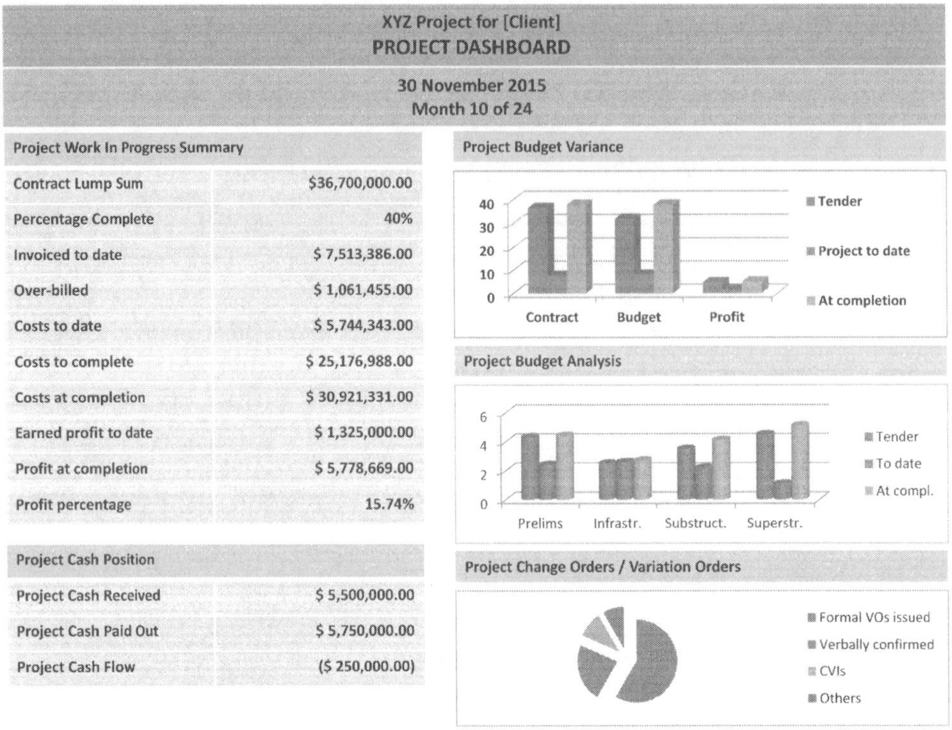
XYZ Project for [Client]
PROJECT DASHBOARD
30 November 2015
Month 10 of 24
Project Work In Progress Summary
Contract Lump Sum $36,700,000.00
Percentage Complete 40%
Invoiced to date $ 7,513,386.00
Over-billed $ 1,061,455.00
Costs to date $ 5,744,343.00
Costs to complete $ 25,176,988.00
Costs at completion $ 30,921,331.00
Earned profit to date $ 1,325,000.00
Profit at completion $ 5,778,669.00
Profit percentage 15.74%
Project Cash Position
Project Cash Received $ 5,500,000.00
Project Cash Paid Out $ 5,750,000.00
Project Cash Flow ($ 250,000.00)
Project Budget Variance
40
30
20
10
0
Contract
Budget
Profit
Tender
Project to date
At completion
Project Budget Analysis
6
4
2
0
Prelims
Infrastr.
Substruct.
Superstr.
Tender
To date
At compl.
Project Change Orders / Variation Orders
Formal VOs issued
Verbally confirmed
CVIs
Others

Figure 4.2 Contents of a project dashboard.

the prudent design professional will pay close attention to the detailed provisions of their services contract and if site visits are not to be made, all contracts (including the owner–contractor agreement), should be revised accordingly. Where even limited site services are contemplated, the agreement should precisely describe the limited services and the frequency of the visits.

Therefore, the contract terms may not exonerate the design professional in all situations and any legal analysis will always begin with an examination of the contract terms and the contract language. For that reason, architects and engineers should review contract provisions regarding inspection and observation duties with legal experts to ensure the language does not place undue risk on the design professional.

4.5 Project Progress Meetings

During the construction stage, the PM / engineer will hold regular formal Project Progress Meetings (normally monthly) attended by the contractor(s) and all relevant consultants, plus the employer or employer's representative. As these meetings will be required to make decisions affecting the project, it is important that all attendees have sufficient seniority within their own organisations to make these decisions; otherwise valuable time will be lost as they go back to their own organisations for authority, potentially leading to project delays.

Project Progress Meetings are an opportunity to:

- a receive progress reports from the contractor, which are often a synopsis of their own internal progress meetings with sub-contractors;
- b receive progress and cost reports from the project consultants;
- c receive records of all labour on site;
- d receive progress photographs – taken by contractor and client if appropriate;
- e discuss actual progress against planned progress;
- f approve appropriate testing procedures and schedules;
- g approve any mock-ups, if necessary, including off-site fabrications;
- h discuss any quality and HSE issues;
- i record weather reports for the project location;
- j highlight any issues with neighbours, such as noise, vibration, right to light, access, safety etc.
- k look ahead to the next period, if there are any issues which could affect progress.

Minutes of the meetings should be prepared immediately after each meeting and should be signed and accepted by all parties, so that it becomes a true record of progress on site. This true record is vital if there are any disputes or disagreements later in the project, especially regarding delays.

Table 4.4 gives a standard comprehensive agenda for a construction Project Progress Meeting.

4.6 Updating the project programme / schedule

When the contractor is issued with the Notice to Proceed (NTP), they are normally contractually required to supply the PM / engineer with a construction programme / schedule, termed the Baseline Schedule in many international contracts. Therefore, throughout the construction period, the contractor and PM / engineer will update this baseline schedule to reflect the actual progress on site.

4.6.1 Establishing the initial construction programme / baseline schedule

Depending on the conditions of contract, the contractor will either submit their programme within a stipulated time, or the PM / engineer may schedule a planning meeting to review the requirements of the contract concerning the schedule or to discuss any project-specific issues required for preparation of the schedule. In some cases, it may be preferable for this schedule to be mutually agreed and developed by the PM / engineer and contractor, so that both sides agree and have ownership of the programme, but in the majority of contracts, the contractor will submit their programme for approval / acceptance / endorsement by the PM / engineer: more different terms for effectively the same concept.

Upon receipt of the contractor's progress schedule submittal, the submittal will be reviewed within the time limit allowed in the contract to ensure that the Baseline Schedule is established in a timely manner. As the project progresses through the construction stage, the contractor's progress schedule submittals will be reviewed by the PM / engineer and compared against the initial programme / baseline schedule.

Table 4.4 Comprehensive agenda for a Project Progress Meeting

PROJECT AT [PROJECT NAME]

FOR

[EMPLOYER NAME]

MONTHLY PROJECT PROGRESS MEETING

INTRODUCTION AND GENERAL ISSUES

1 All project meetings start with acknowledgement of the parties present and a short welcome by the PM.

2 Corrections to / Approval of minutes of last progress meeting.

3 Outstanding issues – see suggested log of outstanding issues below.

Outstanding Issues Log						
Initial Date	Resolution Date	Description	Required Action	Action by	Time of Resolution	
					Planned	Actual

4 HSE Issues

CONSTRUCTION PROGRESS AND TIME ISSUES

5 Work in Progress
- a Main contractor's work (to date and look ahead)
- b Sub-contractor's work (to date and look ahead)

6 Progress Schedule
- a Current schedule update and milestone summary
- b Critical path activities
- c Current expected completion date
- d Manpower and resources issues (planned v. actual, Actual this period, Cumulative actual, variance and justification)
- e Next schedule update

7 Critical delays (i.e. items on critical path)
- a Any delays since last meeting
- b Current known delays (excusable / non-excusable / concurrent)
- c Potential future delays (excusable / non-excusable / concurrent)

8 Non-critical delays (i.e. items NOT on critical path)
- a Any delays since last meeting
- b Current known delays / Potential future delays

9 Extensions of time
- a Requested by contractor
- b Compensable (already granted – pending assessment)
- c Non-compensable (already granted – pending assessment)

SCOPE OF WORK ISSUES

10 Testing of materials and equipment
- a Tests conducted including results
- b Non-conformance reports including corrective actions required

11 Shop drawings and submittals (if appropriate)
- a Under review
- b To be submitted (including dates)

12 Environmental issues
- a Regulated materials including waste
- b Work permits and licences

13 Variation order status
- a Submitted, reviewed, approved, rejected, pending

14 Requests for Information (RFI)
- a Outstanding, expected

COMMERCIAL / FINANCIAL ISSUES

15 Estimates and Cost Projections

16 Contractor payments
- a Submitted, reviewed, approved, returned, pending

17 Value engineering change proposals
- a Submitted, approved, rejected, pending

18 Agreement on final quantities and rates

DISPUTES

19 Items in dispute

20 Claims – submitted, reviewed, accepted, rejected, referred

For this reason, the initial programme is also referred to as the Schedule of Record (SOR) which acts as a benchmark against which any claims for delay, disruption and extensions of time will be compared. It is also useful for establishing the control systems for the purpose of monitoring and assessing progress, managing the day-to-day operations and for coordinating all work required to complete the project.

4.6.2 Monitoring the work and assessing progress

As soon as the project gets underway, accurate as-built schedule information will be recorded to create a historical record of the project. The as-built schedule information will be used for reviewing the contractor's monthly progress update schedules for accuracy and for performing periodic schedule analysis to identify any deviations from scheduled performance to determine if and when corrective actions are necessary for timely completion of the project. The as-built information will also be used to update the schedule when carrying out any impact analysis in order to evaluate the effects of variations and other time-related changes in the work or work plan.

If required by the contract, the contractor will supply a weekly or monthly schedule update of the project to accurately reflect the current status of the work as well as to represent the contractor's proposed plan to complete the remaining work. The PM / engineer should also ensure that the contractor's progress schedule update submittal is reviewed within the appropriate contractual time limits. Following this review and acceptance, the updated schedule will be used to monitor, coordinate and assess progress of the work and serve as the baseline for future schedule impact assessments. If the actual progress of the works has slipped behind schedule, the PM / engineer will need to put into place a strategy for bringing the progress back onto schedule – which is not the same as 'acceleration', as this term is normally defined as speeding up the project in order to finish early.

4.6.3 Maintaining schedule control with Variation and Change Orders

When variations and changes occur which have a time impact, the baseline schedule will need to be amended and the revised schedule approved by the PM / engineer. Most standard forms of contract allow the contractor to make a proposal for the cost and time impact of the change / variation and will do this by carrying out a schedule impact assessment (SIA). This is also the normal procedure even when the contract states that the PM / engineer should evaluate the change impact. A rigorously conducted SIA will demonstrate any impacts to the contractor's work plan or schedule and may also substantiate any requests for adjustments to the contract for such time-related changes.

The baseline schedule will therefore be revised to reflect any schedule impacts for variations and changes to the work or work plan. The following describe the various conditions under which the schedule may be revised:

1 If the contractor proposes to make any significant changes to their work plan including changes to phasing, sequencing, resources and methods, the contractor is required to submit a revised schedule for review and acceptance by the PM / engineer.
2 If the approved programme no longer reflects the contractor's current work plan, the contractor will be required to submit a revised schedule for review and acceptance.
3 If the engineer authorises any changes to the work that has or will significantly impact the contractor's work plan or schedule, the contractor is required to submit a revised schedule to incorporate these changes.
4 If the works have been significantly impacted by issues beyond control of the contractor (*force majeure*), the contractor should submit a revised schedule to incorporate the changes, and most standard forms of contract allow an extension of time for such issues.

4.6.4 Revising the programme / baseline schedule

The latest programme should be that which is approved by the PM / engineer to reflect any changes in the contractor's work plan or to incorporate any formally issued variations and changes. Upon acceptance by the PM / engineer, the revised programme will be communicated to all parties in the project and will replace the baseline schedule as the current approved programme. The revised programme will be used for all subsequent monitoring of the work, managing the day-to-day operations, and for coordinating all work required to complete the remainder of the project.

The PM / engineer must continue to review and identify any potential concerns that may result in schedule delays, safety hazards, quality issues or possible cost overruns. The potential schedule issues should be addressed early and proactively before any major problems are encountered.

4.7 Construction Completion Report

A Completion Report should be prepared for each individual Works Contract when the Works have been substantially completed and a Taking-Over Certificate has been issued together with an 'Interim Payment Certificate at Completion', if appropriate under the particular contract conditions.

The main body of the report should be prepared, at least in draft, before the senior site supervision staff are demobilised, otherwise the necessary detailed knowledge for report preparation might be lost. The report format should be established and early sections drafted as soon as possible. The report should provide a draft final account for the works plus a copy of the snagging list / punch list which must be carried out during the defects correction period – which should be included in the Taking-Over Certificate anyway. ('Snagging' is the term used for remedying minor defects – snags – and is carried out shortly before substantial completion when the works is considered complete by the contractor and offered for inspection. See Chapter 11, section 11.1.4, for full discussion of both snagging lists and punch lists.)

There may be further defects identified during the Defects Correction Period which the contractor will be instructed to remedy and could influence final costs. In addition, any contractor's claims for which the employer is liable may also affect the final cost. Therefore, a realistic contingency should be added to the estimate of final cost to cover this possibility.

Not until the Defects Correction Period has expired and all claims / disputes have been resolved can the final costs be accurately determined. Therefore in this report the PM can only provide an estimate of the final costs.

See Chapter 11 for a fuller discussion of defects correction, final completion and close-out.

4.8 Performance Reports on Contractors

Many employers, especially those with substantial and ongoing construction programmes, require to maintain performance information on both contractors and service providers which will form part of the prequalification procedures for future contracts and may also be used for direct selection purposes. In the USA, there is a centralised system called CPARS (Contractor Performance Assessment Reporting System) which is a web-enabled application that collects and manages the library of automated reports from individual employers and provides a record, both positive and negative, on a given contractor during a specific period of time.

The Performance Report would include:

- conforming to requirements and to standards of good workmanship;
- forecasting and controlling costs;

Table 4.5 Contents of a Project Completion Report

PROJECT AT [PROJECT NAME]

FOR

[EMPLOYER NAME]

CONSTRUCTION COMPLETION REPORT

A – EXECUTIVE SUMMARY:

- i Name of designer (state whether 'in-house' by client or design consultancy)
- ii Project map
- iii Scope of project including type of work
- iv Working conditions
- v Summary of major issues affecting the project
- vi Cost summary including benchmarks (costs per functional unit)

B – EMPLOYER'S REPRESENTATIVE STAFF AND CONSTRUCTION SUPERVISION

- i Manpower breakdown of supervision staff, including any requirement to report nationality
- ii Vehicles, plant and equipment logs used by supervision staff
- iii Outside workforce (if applicable)

C – CONTRACTOR

- i Contractor's name and address
- ii List of sub-contractors names and addresses
- iii List of manpower and major items of plant and equipment (incl. sub-contractors)
- iv Contractor's performance using project KPIs

D – DESIGN AND CONTRACTOR'S METHOD OF WORK

- i Basis of design including Environmental Impact Assessment (if any)
- ii Construction methods (including any unique or special processes)
- iii Major revisions required (for example, by variations to contract)

E – CONTRACT PROGRESS HISTORY

- i Date contract advertised, awarded and commenced
- ii Dates of any significant project milestones
- iii Dates of any work stoppages, with reasons and consequences
- iv Dates of any partial possessions or sectional completions
- v Dates of substantial completion and final completion of the project
- vi Copies of all interim and final certificates issued by PM / Engineer
- vii Other items affecting schedule (e.g. inclement weather, labour disputes, etc.)
- viii As built construction schedule

F – MATERIALS APPROVALS AND TESTING

- i Log of all materials approval forms and testing certificates

G – CONSTRUCTION CONTRACT

- i Tender documents including pricing information
- ii Form of contract used including any amendments
- iii Special Provisions
- iv Post tender negotiations

H – REVIEW OF SCHEDULE OF QUANTITIES AND COST SUMMARY

i Tender vs. actual quantities (if re-measured) with reasons for any significant differences
ii Additional items not included in tender documents with reasons
iii Final Pre-Tender Estimate (PTE)
iv Contract Lump Sum (if appropriate) or Target Cost
v Summary of approved variations to contract
vi Actual Final Account (Out-turn cost)
vii Percentage difference between Final account and PTE / lump sum, with reasons

I – CONTRACTOR'S MAJOR CLAIMS

i Log of all major claims by contractor with resolutions

J – PHOTOGRAPH INDEX (limit of number of photos)

i Importance is on the 'before' and 'after'
ii Original terrain, at major construction milestones and 'as-built' colour photographs – labelled and dated

- adherence to schedules, including the administrative aspects of performance;
- reasonable and cooperative behaviour and commitment to customer satisfaction;
- integrity and business ethics; and
- business-like concern for the interest of the customer.

Plus any other issues which individual employers feel necessary and would influence their choice of reemploying the contractor in the future.

4.9 Summary and tutorial questions

4.9.1 Summary

In order to perform effectively, the end product needs to be very clearly defined, so that the performance can be managed and measured. For a construction project, this end product will be defined and contained in the contract documents and all efforts will be directed towards completing the project scope. In order to achieve that end product, the contractor delivers and the PM / engineer supervises this delivery. This is therefore the rather simplistic summary of the construction stage of the project.

Taking figure 4.1, the contract documents are therefore the 'Plan'. The way that the contractor 'performs' will be set out in the Method Statement and programme / schedule, which are normally presented to the PM / engineer for their acceptance and / or approval.

Control of the performance throughout the construction stage is undertaken by regular monitoring and supervision by the PM / engineer and other supervision consultants as required. So that all parties are engaged and are able to put into place immediate corrective action if necessary, the following regular meetings and regular formal reports will be produced throughout the period of construction:

a Inception Report
b Kick-off Meeting
c Weekly / Monthly Project Progress Reports
d Weekly Project Progress Meetings
e Construction Completion Report
f Defects Report
g Final Performance Report on Contractor (for future due diligence).

However, the above procedures are all very well if the scope of works does not change, there are no unforeseen external factors affecting the design, time or cost, and everybody is allowed to get on with their jobs diligently and unhindered. Unfortunately, that rarely, if ever, happens. Variations and changes are issued, the ground conditions are not what were expected, there is exceptionally inclement weather etc. Therefore, within the progress procedures outlined above, there must be some allowance for incorporating changes and revising the time and cost of the project. This will then effectively create amendments to the contract documents to be further supervised and monitored as per figure 4.1. Monitoring and progression should therefore be seen as a series of circular loops rather than a linear procedure.

When the project is substantially complete, i.e. the employer or end-user is able to use the building or facility for the purpose for which it was intended, there are bound to be some minor snags to be completed, but the main supervision of construction has been completed and no further changes or variations can be instructed to the contractor. When these snags have been remedied, the contract is complete and a Performance Report may be produced, both on the project itself and the participants – consultants, contractor, sub-contractors and maybe even the employer.

4.9.2 Tutorial questions

1. Outline how the project control loop covers the issues required in construction contract management.
2. What are the main uses of a Project Inception Report?
3. Who should be present at the Project Kick-off Meeting, and why?
4. Why is it vital to maintain full and detailed minutes of Project Progress Meetings?
5. What contemporaneous records should be included as appendices to Project Progress Meetings?
6. If the contractor's programme is not a formal contract document, why should the PM / engineer be required to approve or accept such a programme?
7. Discuss the purposes of a Construction Completion Report.
8. Comment on the major differences in project supervision for a DB contract.

5 Payments to the contractor

5.1 Introduction

As mentioned in Chapter 1, all contracts are an exchange between two parties and in exchange for the contractor using their skills and resources to construct the building / facility, the employer undertakes to pay them. Therefore both parties have benefits and burdens under the contract – see table 5.1.

It is normal practice to agree a figure at the beginning of the contract – i.e. the 'contract sum', which will subsequently be amended due to variations and Change Orders issued by the employer (see Chapter 6), fluctuations (i.e. inflation) and finalising the scope of works / cost of any provisional sums contained in the original contract sum. The term 'lump sum' merely denotes that this is a single figure agreed at the beginning of the project and invariably gives a very spurious level of comfort to the employer, as there many ways why this lump sum will bear no resemblance to the final outturn cost at the end of the project.

Depending on the procurement route chosen, there may not be an actual contract / lump sum at all. Many Design–Build (DB) contracts include a 'target cost', with incentives and penalties should the contractor achieve completion for less than the target cost or, perish the thought, exceed the target figure. In all of these cases, the employer is obliged to pay the contractor for the work which has been properly carried out.

Construction contracts normally fall within the definition of 'entire contracts', which means that consideration (i.e. payment) is given on full performance and completion of the contractor's obligations. Clearly, in a major multi-million dollar project, no contractor will be able to finance the entire contract from mobilisation and commencement all the way up to substantial completion, so provisions for interim payments on account will be written into the contract terms and conditions and such provisions are found in all standard forms of construction contract.

These interim payments can be made in various ways:

a on the basis of work carried out in a given time period (i.e. monthly valuations);
b on the basis of completion of particular elements of work (stage payments or milestones);
c on the basis of what the contractor has paid out in direct costs plus an agreed percentage for the contractor's overheads and profit (reimbursable).

Figure 5.1 shows a general procedure diagram for the payment process and the times for each of these processes will vary according to the particular form of contract being used on the project.

Table 5.1 Benefits and burdens in a construction contract

	Contractor	*Employer*
Benefit	Receive payment	Completed works
Burden	Constructing the works	Make payment

a The contractor will submit a request for payment at the agreed point – i.e. end of month, completion of appropriate milestone etc. This payment request will be generated by the contractor's QS / contract administrator and should only include work which has been technically approved by the supervision consultants.

b The employer's cost consultants will review the request / application to ensure that:

- Only work which has been carried out properly is included, i.e. accepted by the supervision consultants or PM / engineer.
- The pricing of the work is in accordance with the contract documents, especially the pricing document (BOQ or schedule of rates etc.), which will include the appropriate rates agreed at tender stage and quantities as necessary. If work has been included in a payment request which does not have an appropriate rate in the BOQ, a rate should be calculated or otherwise agreed, which may require a formal Variation Order to be issued. The method for calculating such 'star rates' is often set down in the standard form of contract.

c If there are problems with the payment request or it is not compliant with the contract documentation or project procedures, the application / request will be returned to the contractor for amendment. Compliant requests will be forwarded to the PM / engineer for approval or endorsement, depending on the specific procedures for the project.

d As stated in Chapter 1, the PM / engineer normally has the responsibility and authority to 'sign-off' the contractor's interim payments and this authority should be clearly stated in both the main contract terms and conditions as well as the PM / engineer's professional services agreement (PSA). Once approved, the payment application would either go direct to the employer for actual payment to the contractor, or be returned to the contractor with an instruction to submit a formal invoice for the approved amount. The contractor will not wish to submit a formal invoice until the amount has been agreed as any alterations to a submitted invoice may cause difficulties with accounting software packages. It is an unfortunate sign of the times that integrated financial and business software such as SAP and Oracle can be extremely beneficial providing you do things their way. Trying to make ad hoc or post hoc amendments can be excruciatingly painful and time consuming.

e As employers have become much more 'sophisticated' in that they are more knowledgeable regarding the workings of the construction industry – which is especially true of experienced clients such as property companies and developers (see *Introduction to Building Procurement*, Chapter 2); there will invariably be an experienced 'project sponsor' or 'contract holder' who acts as the focal point of the employer with the project team and therefore may also require to endorse or approve the payments before they are actually paid.

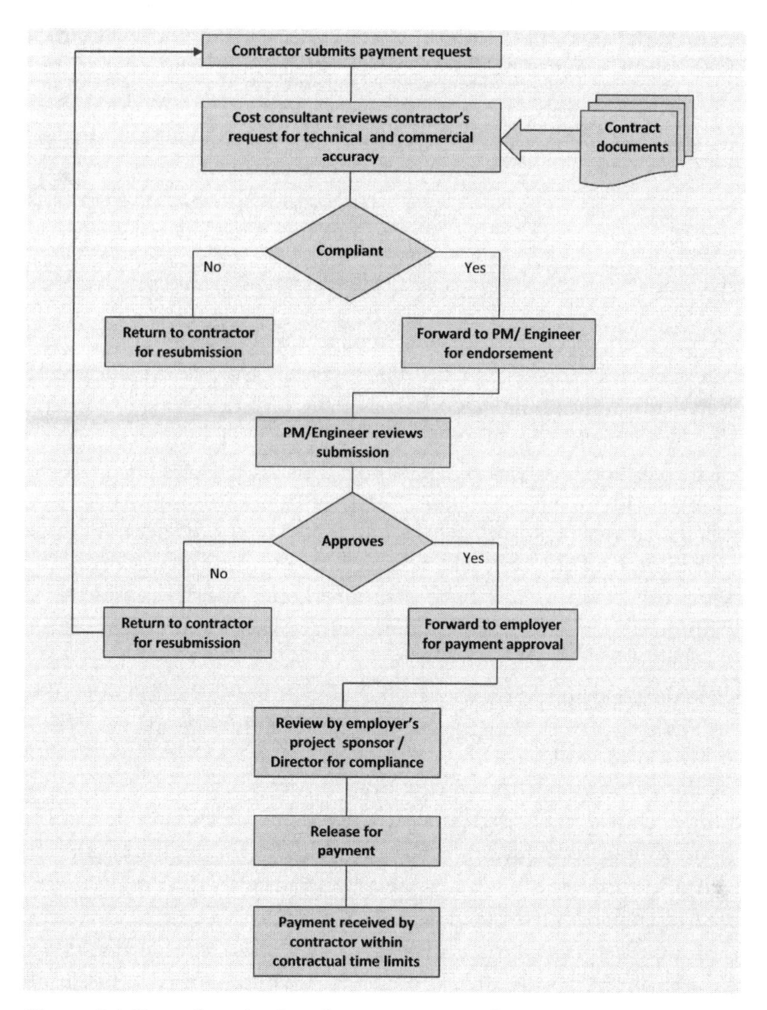

Figure 5.1 Procedure for interim payments to the contractor.

There are several reasons for the requirement for contractors to submit invoices, rather than just receive the 'payments on account' which occurred historically:

- Government authorities require invoices for assessment of corporation tax and VAT.
- Large companies now use integrated accounting packages as mentioned above, therefore formal invoices with specific reference codes are required for corporate financial management and cost control.
- Invoices are more formal and can show a more definite debt owed to the contractor by the employer.

All standard forms of contract will have time limits for each of these stages to be carried out and these may or may not be amended for individual specific projects. Indeed, on many contracts in certain parts of the world, the employer often completely ignores their contractual obligations and pays the contractor as and when the fancy takes them.

Clearly, this is in breach of contract and the contractor is within their rights to take any remedies that are available to them. In doing so, they are taking both a contractual and a commercial decision which may have consequences with future relationships with that and possibly other employers.

Figure 5.2 gives a generalised timeline for processing interim payments in a construction contract.

5.2 Pricing documents

5.2.1 Bills of Quantities

Bills of Quantities (BOQs or BQs) have existed in one form or another for over 300 years and serve to itemise the finished work in a construction project with standard descriptions which are mutually understood by both the BOQ producer (usually consultant QS) and the BOQ user (contractors). This is achieved by the use of clearly understood rules of measurement or standard methods of measurements (SMMs – see below). As stated above, the BOQ is usually prepared by a consultant quantity surveyor engaged by the employer (often known as the PQS), based on detailed drawings and specifications prepared by the project architect / designers. In the post-contract or construction stage, the BOQ assists both the contractor and the employer's cost consultant in the valuing of progress payments and variations and provides a financial structure for commercial contract administration.

A BOQ can be prepared using various alternative standard methods of measurement depending on the nature of the project and its complexity. In the UK, from January 2013, the SMM for building work has been superseded by the 'New Rules of Measurement' (NRM), although for infrastructure projects, the Civil Engineering Standard Method of Measurement – 3rd edition (CESMM3) is still in common use.

There are also other specific methods of measurement for highway and bridge works and some larger employers have also developed their own methods usually based on one of the above main methods.

Internationally, many countries have developed their own standard methods of measurement, e.g. Ireland, Australia and Malaysia, which are more appropriate for the particular regulatory conditions in their country as well as different construction techniques and materials.

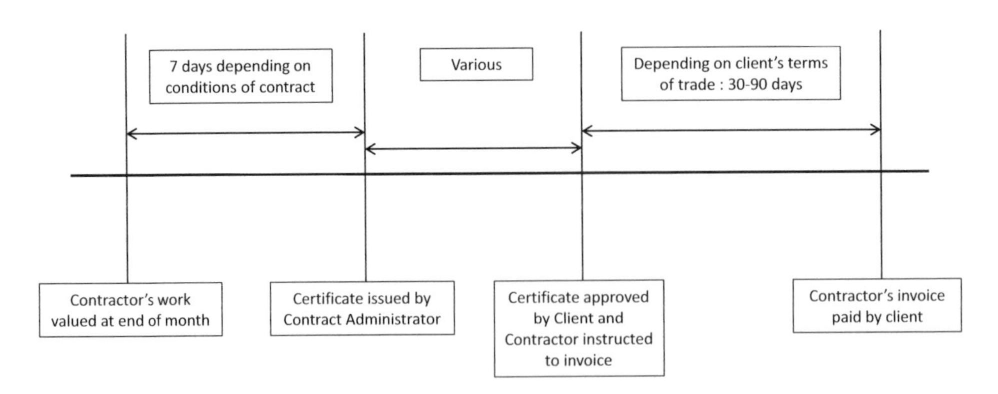

Figure 5.2 Timeline for processing Interim Payment Certificates.

The contractual status of the BOQ can vary in that it can form part of the contract documents with the quantities being considered firm or it could be provided as an approximate Bill of Quantities which require re-measurement during or after construction. In the latter case, the approximate BOQ would not form part of the contract documents.

The production of a full BOQ 'taken off' from a fully detailed design requires considerable time to prepare during the latter stages of the design phase and should result in more competitive and accurate tenders with fewer ambiguities and therefore less opportunity for disputes during the construction stage of the project. The process of producing a BOQ, however, requires the PQS to interrogate the design and specification in considerable detail, which also enables them to identify inaccuracies and inconsistencies in the drawings and specification prior to tender, which also helps to further reduce any subsequent post-contract problems.

Advantages and disadvantages of Bills of Quantities during the construction stage

The advantages are as follows:

1 *Calculation of interim valuations and progress payments* – The rates are used to value the work completed to date during the construction stage and make progress payments to the contractor.
2 *Valuing of variations / Change Orders* – The rates in the BOQ are used for valuing variations and changes which have been authorised by the project manager, whether as additions or deductions. Therefore the variations are valued on the same basis as the original tender.
3 *Assessing the final account* – The final cost of the work will be based on the rates in the BOQ.
4 *Asset management* – The BOQ provided readily available data for asset management of the completed building, life cycle costing studies, maintenance schedules, general insurance and insurance replacement costs.
5 *Taxation* – BOQs provide a basis for quick and accurate preparation of depreciation schedules as part of a complete asset management plan for the project.

The disadvantages, on the other hand include:

1 *Estimating practice* – Tenderers may ignore the formal specification document by pricing only according to the BOQ. This may lead to under-pricing and the consequent risk of unsatisfactory performance during the construction stage.
2 *Procurement* – The use of a detailed design and associated BOQ may discourage contractors from submitting alternative design solutions and value engineering change proposals (VECP). A firm BOQ is only suitable to the traditional procurement system.
3 *BOQ errors* – Because the BOQ is a complex document and developed from a design which may not be 10 per cent complete, there are likely to be errors, omissions and discrepancies between the drawings, specification and BOQ. The contract should make it clear which document has priority.

Structure of Bills of Quantities

Historically, Bills of Quantities have been structured in accordance with the standard method of measurement which has been used for the project. In the UK, this has been the latest version of the SMM published by the RICS. As stated above, from January 2013, SMM edition 7 was replaced by the New Rules of Measurement (NRM) and NRM2 relates to the detailed measurement of building works – NRM1 relates to cost estimating and cost planning during the design stage.

The Introduction to NRM2 states:

> This volume provides fundamental guidance on the detailed measurement and description of building works for the purpose of obtaining a tender price. The rules address all aspects of bill of quantities (BQ) production, including setting out the information required from the employer and other construction consultants to enable a BQ to be prepared, as well as dealing with the quantification of non-measurable work items, contractor designed works and risks. Guidance is also provided on the content, structure and format of BQ, as well as the benefits and uses of BQ.
>
> While written mainly for the preparation of bills of quantities, quantified schedules of works and quantified work schedules, the rules will be invaluable when designing and developing standard or bespoke schedules of rates.
>
> These rules provide essential guidance to all those involved in the preparation of bill of quantities, as well as those who wish to be better informed about the purpose, use and benefits of bill of quantities.

Bills of Quantities produced in accordance with the NRM will normally comprise the following sections:

- form of tender (including Certificate of Bona Fide Tender)
- summary (or main summary)
- preliminaries, comprising two sections as follows:
  - information and requirements; and
  - pricing schedule
- measured work (incorporating contractor designed works)
- risks
- provisional sums
- credits (for materials arising from the works)
- dayworks (provisional)
- annexes.

When using NRM, the structure will generally be in accordance with the tabulated work sections given in the document:

1 Preliminaries
2 Off-site manufactured materials, components and buildings
3 Demolitions
4 Alterations, repairs and conservation
5 Excavating and filling

6 Ground remediation and soil stabilisation
7 Piling
8 Underpinning
9 Diaphragm walls and embedded retaining walls
10 Crib walls, gabions and reinforced earth
11 In-situ concrete works
12 Precast / composite concrete
13 Precast concrete
14 Masonry
15 Structural metalwork
16 Carpentry
17 Sheet roof coverings
18 Tile and slate roof and wall coverings
19 Waterproofing
20 Proprietary linings and partitions
21 Cladding and covering
22 General joinery
23 Windows, screens and lights
24 Doors, shutters and hatches
25 Stairs, walkways and balustrades
26 Metalwork
27 Glazing
28 Floor, wall, ceiling and roof finishes
29 Decoration
30 Suspended ceilings
31 Insulation, fire stopping and fire protection
32 Furniture, fittings and equipment
33 Drainage above ground
34 Drainage below ground
35 Site works
36 Fencing
37 Soft landscaping
38 Mechanical services
39 Electrical services
40 Transportation
41 Builder's work in connection with mechanical, electrical and transportation installations.

These work sections reflect the structure and nature of the modern construction industry better than the predecessor SMMs and also match the structure of the RIBA Plan of Work and OGC Guidance Notes.

Preliminaries

These are the general items usually associated with the contractor's site establishment on the project. A look through the relevant section of the NRM shows various information requirements and general cost items included in the project. As preliminaries are effectively site-based overheads, their costs are not related directly to the quantity of work but rather the duration of the project (termed time related charges) and the method adopted to construct the works (termed method related charges). Preliminary items fall under the following headings:

- General Employer's Requirements
- Security, Safety and Protection of the Works
- Limitations on Method, Sequencing or Timing
- Temporary works and services
- Contractor's Management and Staff
- Site Accommodation
- Contractor's mechanical plant
- Any works or materials supplied by employer.

The contractor will normally price the preliminary items as a lump sum, but in some cases they are required to price the item as a time related charge or a fixed charge for a particular event. For example, the major cost of providing tower cranes will be (a) erecting the crane, (b) rental for the period it is on site and (c) dismantling the crane; none of these costs relate to the amount of work it does but to either an event or a time period. Therefore, when payment is made to the contractor in Interim Certificates, the valuation of these items must take this into account.

Prime cost and provisional sums

Prime Cost Sums (PC Sums), which have now disappeared from use in the UK but are still in use in some international jurisdictions, are a procedure to include the cost of a nominated sub-contractor or a nominated supplier into the main contractor's tender figure and contract sum. Their costs, which acted as a prime cost to the contractor (i.e. the equivalent of labour, plant and materials) would be covered by the PC Sum and the main contractor was entitled to add a percentage profit to this sum and also allow for general or special attendances on the sub-contractor (i.e. provision of welfare facilities, health and safety responsibilities, power supplies etc.). When the nominated sub-contractor's final account was received, this replaced the PC Sum in the main contractor's final account.

Provisional Sums are included in the BOQ for work which is not fully designed at tender stage but an allowance is required in the contract sum, for the purposes of the client's overall project budgeting and also to ensure the contractor is aware of the extent of the works. Contingencies are included in project costs as a provisional sum. Provisional sums have now taken over the role of the PC Sum, in that organisations which would have been covered by a PC Sum are now covered by a provisional sum, such as statutory undertakings – those organisations who are the only ones allowed to do certain work, such as connecting to the mains electricity, mains gas etc.

As with SMM7, NRM2 allows provisional sums to be stated as defined or undefined. Defined provisional sums mean that the contractor has included the scope of the work as part of their programme and therefore cannot claim an extension of time or loss and expense as a result of the architect or contract administrator firming up the actual scope, even if it increases the cost of the sum. Undefined provisional sums on the other hand mean that the contractor has not included the extent of the works in their programme and therefore may be able to claim for additional time or cost.

Dayworks may be included as a provisional sum within the overall contract sum or can be totally excluded from the contract sum with the contractor's daywork rates itemised separately. If the latter is the case, then clearly care must be taken at tender evaluation stage to ensure that the rates are fair and reasonable, since they do not contribute to the overall contract sum comparison. Dayworks should only be used to value work where no other method is appropriate, i.e. as a last resort, although many contractors are keen to be paid on dayworks as this work is effectively valued at cost plus – and a very nice cost plus thank you very much.

5.2.2 Bills of Approximate Quantities and Schedules of Rates

Bills of Approximate Quantities and Schedules of Rates are structured and prepared in exactly the same way as a firm BOQ, except that with the Bill of Approximate Quantities, the quantities are not guaranteed and will be subject to re-measurement when the work is carried out on site. The purpose of putting an approximate quantity in the tender documents is to give the contractor an indication of the extent of the item required, so that any economies of scale can be calculated. Clearly, the unit rate (in £/m^3) will be lower if the project requires 1000 m^3 than if it only requires 5m^3. Civil engineering projects prepared under the ICE Conditions of Contract are 'remeasurement contracts', meaning that the quantities in the BOQ are not guaranteed and must be measured separately when the work is carried out on site, therefore not only do the BOQs only give an indication of the scope of the item, but in the concrete work section of CESMM3, the contractor is also given an indication of the total amount of concrete across all items.

A Schedule of Rates goes one step further in that there are no quantities included and the tendering contractor is expected to guarantee a rate which will be used if there are 1000 m^3 or only 5m^3 required. Clearly this does not allow for any economies of scale to be given and should only be used for relatively small projects or as part of a term contract where the contractor is required to carry out work at very short notice.

5.2.3 Activity Schedules

An Activity Schedule is a list of the activities which the contractor expects to carry out in completing their obligations on the project. It is only relatively recently that this has been used as a pricing document and clearly there are advantages in using a time / programme-based payment mechanism rather than a quantity-based mechanism as with the BOQ. When the Activity Schedule has been priced by the contractor, the sum for each activity or each group of activities is the price to be paid by the employer for that activity or group. The total of all the activities and groups is the contractor's total price for carrying out the works (i.e. the contract sum).

It is essential that the activity descriptions are clear, unambiguous and complete so the entire works are included within the overall activity descriptions and the work included within any one particular description can be readily identified. Since payment is usually only made by the client on completion of each activity or group and not before, each activity description should include a definition of the measure to be adopted to confirm completion – e.g. signing off by the project manager or contract administrator.

This form of payment mechanism is adopted in many standard forms of contract as it can significantly reduce the administration burden. It is used particularly in DB contracts where the contractor has control over the definition of the project but can also be used effectively in more traditional procurement routes.

The NEC3 Contract uses an Activity Schedule as the payment mechanism in Options A and C and internationally, the FIDIC Yellow Book (for Plant and Design–Build projects) and Silver Book (for EPC / Turnkey projects) at Clause 14.4, allows payments to be made by instalments against a Schedule of Payments, which may be defined by reference to actual progress. This would therefore allow an Activity Schedule to be adopted.

See figure 5.3 below for an example of an Activity Schedule giving the contractor opportunity for pricing the scheduled activities.

5.3 Retention

It has always been the custom and practice in the construction industry for the employer to withhold a proportion of the gross valuation of work carried out in each Interim Payment Certificate. This is termed retention and normally represents 5 per cent of the gross valuation certified to date. The purpose of a retention fund is to give some comfort to the employer, in case the contractor fails to perform the contract fully or fails to rectify defects when instructed to do so following practical or substantial completion of the works.

If retention is to be included in a contract, the actual percentage must be set out in the contract particulars together with how this will be returned to the contractor on completion. Normally, half of the accrued retention fund is returned on the issue of the Certificate of Practical / Substantial Completion and the other half returned following final completion of the works.

The money in the retention fund rightfully belongs to the contractor – they have earned it and it has been certified in the gross valuation of work carried out. Under normal circumstances, the monies will be kept by the employer and are therefore under their control but the employer must behave as though they were a trustee for the contractor, except that they have no obligation to invest the monies and will only return the precise sums which they originally retained.

So what happens if, or when, there is a dispute and the employer considers they have a right of set-off against the contractor? This will depend on whether the set-off has been independently verified, for example, in an adjudication. If the employer's claims are speculative and unsubstantiated, the contractor may ask for the retention fund to be held in a separate account (called an escrow account) to which the employer has no direct access. In fact, many contractors have justifiably queried why the employer's working capital should benefit from their monies when there are other ways of giving the same degree of comfort to the employer.

One of the other ways is a retention bond which the contractor can purchase from a bank or other financial surety for the appropriate proportion of the gross valuation. In this way, the employer may draw on the bond, should they so need, which will act in the same way as the retention fund explained above but without affecting the contractor's cash flow. Additionally, if the contractor is required to submit a

	Task Name	Duration	Preceded	Labour	Material	Plant	TOTAL	Aug					Sep				Oct				Nov			Dec					
			by	cost	cost	cost	COST	3	10	17	24	31	7	14	21	28	4	12	19	26	2	9	16	23	30	7	14	21	28
1	Demolition	6 days																											
2	Site Preparation	6 days	1																										
3	RC piles	20 days	2																										
4	Drainage works	20 days	2																										
5	Excavation & Support	30 days	4																										
6	Raft Foundation	6 days	5																										
7	Steel frame	21 days	6																										
8	Roof covering	6 days	7																										
9	External wall cladding	20 days	7																										
10	Internal walls & ceilings	25 days	7																										
11	Power supply system	30 days	10																										
12	Lighting system	15 days	10																										
13	HVAC	15 days	10																										
14	IT network	10 days	10																										
15	Internal finishings	15 days	10																										
16	External Works	21 days	9																										
17	Final cleaning & Handover	3 days	all																										

Figure 5.3 An example of a Priced Activity Schedule.

performance bond, this will also give comfort to the employer that the obligations will be fulfilled. Needless to say, on many contracts the contractor is required to submit a performance bond as well as being subject to retention.

5.4 Advance payment

Advance payment provisions are used in construction contracts to assist the cash flow of the contractor at the early mobilisation stage when significant costs can be incurred in setting up the site, bringing in large items of plant and equipment etc. Therefore having an advance payment means that the contractor does not have to pay for all of this mobilisation from their own funds.

Advance payments are common in international contracts where the costs of mobilisation are very high and should be paid as an interest-free loan from the employer to the contractor. The advance payment can be paid in one or more instalments depending on the nature of the project and the lead time required for mobilisation of plant, equipment and people. All such details should be set out in the contract, so that there is no misunderstanding by the parties regarding how much and when the payments are to be made. The payments will be made via a standard Interim Payment Certificate which will be gradually repaid from valuations of work done over the course of the contract.

5.4.1 Advance payment bond / guarantee

In order to protect the employer from contractor insolvency before the advance payment has been fully repaid, the employer will normally require the contractor to provide an advance payment bond or guarantee which will protect the employer if the contractor fails to fulfil their obligations under the contract.

An advance payment bond will normally be an on-demand bond, meaning that the bank issuing the bond is obliged to pay the amount of money set out in the bond immediately on demand, without any preconditions having to be met. This is as opposed to a conditional bond where the bank is only liable if it has been fully established that there has been a breach of the contract. Advance payment bonds must clearly be very carefully drafted to ensure that all terms and conditions are crystal clear and understood by both parties as well as the issuing bank.

A more complicated issue regarding advance payments is what happens if the contract is terminated by the employer before the advance payment has been fully repaid. A situation may well arise where the contractor has effectively been overpaid because of the advance payment. If a bond has been provided, it would be wise to state clearly how such a bond is to operate if the contract is terminated in circumstances where there is an existing claim to an advance payment.

5.5 Pay Less Notice / Withholding Notice

In the good old days, when employers knew little or nothing about the construction industry or its machinations, the PM / engineer and cost consultants would certify payment and the employer would pay – invariably without question. As employers became increasingly more sophisticated, they began to question the amounts certified by the cost consultant and / or PM / engineer to the extent that they may even reduce the payment to the contractor (termed withholding). Over time, this became quite a sore point and as procedures changed to require the contractor to submit their valuation first (instead of cost consultant and contractor valuing the work jointly), there was a significant amount of reductions in what the contractor felt was due. This of course assumes that the contractor had not inflated their original valuation!

Something clearly had to be done about this on a formal basis, and in the UK, the Construction Act in 1996 allowed a procedure known as 'Withholding Notices' to inform the contractor that they would be paid less than they thought they were going to be paid. This didn't fully work in practice – partly because abatements (i.e. reducing the contractor's valuation due to defective work) did not require a Withholding Notice – and was subsequently amended by the 'New Construction Act 2009' and the term 'Pay Less Notices' entered the construction industry lexicon.

The timing of the Pay Less Notice is crucial, as if it is not served on time, the notice will not be effective and the payee will be entitled to adjudication or other dispute resolution procedures if the valuation is subsequently reduced.

5.5.1 The content of Pay Less Notices

Under the 1996 Act, a Withholding Notice simply needs to identify the amount or amounts being withheld. However, the 2009 Act requires much more detail with supporting calculations. The grounds for a Pay Less Notice are likely to be one or more of the following:

1 Abatement (reduction) of the value of the work; for example

   a Work not carried out
   b Percentage of work complete not correct / value of work overstated
   c Work not carried out properly / carried out defectively

2 Set-off or counterclaim against the value of the work

   a Work carried out by others
   b Claims by other contractors against the payee contractor

The notice is required to state the amount attributable to each of the grounds above, together with the sum the paying party considers due and how this sum has been calculated.

One of the reasons for revising this aspect of the Construction Act was the failure of contract administrators to issue Payment Notices under the contract. The new Act gives the contractor power to issue their own notice. The paying party under a construction contract (i.e. employer) must understand the statutory requirements for giving a Pay Less Notice if there is disagreement with the amount being requested by the contractor. Failure to comply with the process will mean the notice is void. All notices must therefore be served within the deadlines stated in the contract. If no deadlines are expressly agreed, then (in the UK) the Scheme for Construction Contracts will apply. As indicated above, the contractor can now issue a notice of default specifying the sum it considers due if the employer misses a Payment Notice deadline. Contract administrators take note.

5.6 Payment for materials not yet incorporated into the works

5.6.1 Materials on site

Many standard forms of contract allow the contractor to be paid for materials stored on site (or adjacent to the site in, for example, a specific storage or lay-down area) which are intended to be incorporated into the works. Some engineering type contracts also allow items of plant to be included if the item of plant is itemised and costed separately. The reason for this is again to assist the contractor's cash flow as the materials will need to be bought and paid for significantly in advance of being incorporated into the works – and therefore recouped in Interim Valuation Certificates.

Where the contract allows for materials stored on site to be included in an interim payment, the contractor is required to:

a keep full records of orders, receipts, delivery notes etc. of all materials and goods stored on site;
b maintain full insurance of the materials and goods, whilst in storage;
c fully secure the storage against loss, theft, damage or deterioration.

Payment for materials on site is therefore another form of advance payment and so will not increase the employer's overall costs, as the materials costs are included in the contract sum anyway – in fact, the overall costs may be reduced if the contractor can purchase in advance and negotiate more favourable rates. What is saved is programme efficiency as the materials will be ready and available for incorporation into the works when required by the programme schedule – Just-in-time procedures are all very well in fully controlled environments, but do not work particularly well in the construction industry.

Materials on site generally don't affect the contractor's cash flow forecast, unless more materials are brought on than envisaged, which of course may be a cause of concern in relation to contractor insolvency. The PM / engineer should not allow payment to be made for materials brought onto site too far in advance just for the purpose of boosting the contractor's cash flow. A cash flow forecast should be carried out on the assumption that materials will be brought onto site as and when required unless otherwise stated. An exception to this will be in large process engineering plants such as oil and gas installations where the lead times of large components can be very long indeed; there are two ways of dealing with this, firstly, to give an advance payment to the contractor for the component and secondly, for it to be purchased directly by the client and given as a 'free issue' to the contractor. This second option is common in EPC contracts, especially where the employer has a significant in-house procurement capability.

5.6.2 Materials off site

Materials off site can be an even more complicated subject and can have a significant impact on the cash flow forecast of both contractor and employer. Very few standard forms of contract allow for payment of materials stored off site, as the employer or PM / engineer has little or no control over the off-site storage facilities and there may also be major issues regarding title to the goods if the contractor becomes insolvent.

Materials paid for and stored on site can be secured by the employer in the event of the contractor's default. This may not be the case with off-site storage facilities.

5.6.3 Pre-fabricated components off site

An example is the increasing use of 'pod construction' for student accommodation, budget roadside hotels etc. Under the normal rules of payment, the costs of the pods would be payable as the pods become incorporated into the works. However, as the pods are pre-fabricated off site and therefore a significant amount of the value added and 'construction' is carried out in a factory somewhere away from the site, payment should be made as the pre-fabrication progresses in the factory, not on the site. This can significantly shift the emphasis of the cash flow forecast and reduce the overall construction time (as it is intended to do). Traditionally, the internal finishes and second fix plumbing may not have been completed until near the end of the project whereas with this example payment for the pods may become due during the middle part of the contract as the pre-fabricated pods are completed in a factory controlled environment off site.

It should be noted that some legal jurisdictions do not recognise payment for materials off site, therefore the mechanism to cover such items would be the use of a separate contract of purchase by the employer direct with the materials or component providers and taking care to ensure that there is full design coordination as the contractor is not responsible for this portion and would be given the materials or components as 'free issue' by the employer.

Figure 5.4 shows an example of an Interim Payment Certificate with the necessary calculations.

5.7 Monitoring spending and financial commitment

5.7.1 The need to monitor spending

During the execution and delivery of a construction project, robust and effective procedures for financial control are imperative to ensure that the project does not overspend and also to assess the progress of spending against the total budget; this can also be compared with the physical progress of the scope of works.

Project financial control procedures are primarily intended to identify deviations from the agreed project plan rather than to suggest possible areas for cost savings – that will be discussed in Chapter 8 on value engineering. As discussed in that chapter, the time at which major cost savings can be achieved is during the planning and design stage of the project; during the actual construction, any changes to the design will only serve to delay the project and lead to cost increases. Therefore, the focus of project control and monitoring should be on overseeing that the agreed contract scope is being fulfilled within the agreed budget and schedule, rather than on searching for significant improvements and cost savings. It is only when a firefighting operation is required that major changes will occur in the construction plan, although these can occur more often than they need to.

5.7.2 The employer's project budget

The employer's project budget clearly needs to include considerably more than just the construction costs. Depending on the nature of the project, the project budget will normally be made up of four main areas – acquisition costs, construction and fit-out costs,

EMPLOYER NAME

PAYMENT CERTIFICATE NO.

Contact Description: PROJECT NAME

Contact NO:		Budget Ref
Client:	EMPLOYER NAME	
Project Management Consultant :	PM NAME	
Construction Supervision:	CMCS NAME	
QS Consultant:	QS NAME	
Contractor:	CONTRACTOR NAME	

ORIGINAL CONTRACT PRICE			0.00
APPROVED V.O (NET AMOUNT)	TO V.O NO	0.00	
			0.00
ADJUSTED CONTRACT PRICE			0.00

CERTIFICATE NO.
DATE: XX.XX.XXXX PERIOD ENDING XX.XX.XXXX PAYABLE BY:

VALUE OF WORKS COMPLETED TO DATE		0.00
DISCOUNT @	0.0%	0.00
	SUB TOTAL	0.00
WORKS CARRIED UNDER VARIATION ORDERS		0.00
VALUE OF MATERIALS ON SITE		0.00
TOTAL VALUE OF WORK DONE		0.00

VALUE OF WORKS COMPLETED THIS PERIOD

ADVANCE PAYMENT	X%	0.00
RECOVERY OF ADVANCES		0.00
UNRECOVERED ADVANCES		0.00
PREVIOUSLY RECOVERED		0.00
RECOVERY OF ADVANCES FOR THIS MONTH		0.00

GROSS VALUE CERTIFIED BY THIS CERTIFICATE		0.00
LESS RETENTION	X%	0.00
LESS PREVIOUSLY RECOVERED		0.00
LESS RECOVERY OF ADVANCES FOR THIS MONTH		0.00
PENALITIES / LDs		0.00
OTHER REDUCTION		0.00

NET SUM CERTIFIED BY THIS CERTIFICATE		0.00
RELEASED RETENTION (Previously)		0.00
RETENTION TO BE RELEASED (for this Payment)		0.00
	TOTAL	0.00
NET SUM CERTIFIED BY PREVIOUS CERTIFICATE		0.00
AMOUNT DUE FOR PAYMENT		0.00

RETENTION FROM ABOVE (1)	0.00
RETENTION FROM PREVIOUS CERTIFICATES (2)	0.00
NEV RETENTION (For this Certificate) (3) = (1-2)	0.00
RELEASED RETENTION (Previously & This Payment) (4)	0.00
BALANCE RETENTION TO BE RELEASED (5) = (1-4)	0.00

Constructor PM	Consultant	Construction Supervision	PMC	Project Manager	Head of Engineering Section	Head of Contractor Section	Director
Date	Date	Date	Date	Date	Date	Date	Date

Figure 5.4 Example of an Interim Payment Certificate.

finance charges, developer's profit. In total, all of these costs will be compared to the value of the development (once it is finished) and if the value is greater than the cost, then bingo, we have lift off.

These four main areas of the employer's budget will normally contain the following:

Acquisition costs

- actual purchase costs or rental costs of the site
- real estate agent's fees
- legal costs of land acquisition
- statutory costs of land acquisition
- security and protection of site prior to construction
- insurance.

Construction and fit-out costs

- site survey costs including geotechnical and environmental impact assessments if required
- demolition costs of any existing structures on site
- site preparation, remediation and decontamination work to ground conditions (especially for brownfield sites)
- any enabling works before substantive construction of building / facility
- construction costs of new building / facility, including all labour, materials, plant and equipment
- renovation costs of any existing buildings
- professional fees and expenses – architecture and engineering design, cost consultancy, project management, inspection, supervision, legal etc.
- statutory fees for planning permission, building permits, building control and inspections etc.
- insurance of site and works, if not included with construction costs
- furnishings, fittings and equipment to be installed or included in finished facility
- construction contingencies
- allowance for increased costs / inflation.

Finance charges

- cost of borrowing / finance charges for construction activities
- any bonds, warranties and guarantees required for the project
- VAT or other development taxes.

Developer's profit

- profit element for project investors.

The above list, whilst trying to cover the majority of costs that a developer or employer would incur, clearly cannot be exhaustive and other costs may arise depending on the particular project circumstances, location and size.

As stated above, the other side of the equation from the development costs is the development value, which is either a balance sheet figure (shown as an increase in fixed assets as the new building will increase the property value of the employer's business), or real money income generated by sales or rent of whatever has been built. If it's the latter, then the cash flow of the income stream (i.e. annual rents) will have to be calculated at present day values (called net present value or NPV). This can be compared like-to-like with the costs paid out for construction. The people who do these calculations think it's all very clever and blind us with terms like internal rate of return, year's purchase, yield rates and amortisation. It all boils down to cash going out and cash coming in.

5.7.3 The contractor's project budget

Included within the lists above was the line:

- Construction costs of new building / facility, including all labour, materials, plant and equipment.

Plus a few other issues which usually find their way into the construction project budget, depending on how the different contracts are put together:

- Furnishings, fittings and equipment to be installed or included in finished facility
- Construction contingencies
- Allowance for increased costs / inflation.

Whereas the employer's project budget tends to be lump sums for each of the bullet point costs itemised above, the contractor's costs will be far more detailed and consist more of prime costs, with the exception of those costs payable to sub-contractors and component suppliers. Of course, this will also depend on the procurement route chosen, in that management contractors will also have lump sum payments to their work package contractors – but somewhere down the food chain, the prime costs of labour, materials, plant and equipment will have to be established, budgeted for, paid and controlled.

Figure 5.5 shows a generalised contractor's budget form developed from the Activity Schedule shown in figure 5.3. The tasks or operations will be developed into a work breakdown structure (WBS), the estimated costs of which will make up the initial budget price, and possibly the tender price / contract sum. The actual costs will be established when the work takes place and a difference (or variance) established which will then show if the contractor has increased the profit level or decreased. As most contractors nowadays tender at very low profit margins or even at a loss, increasing the profit margins in ways such as this is crucial to their survival.

5.7.4 Control of expenditure against the budget

Budgets, therefore, are an estimate of both income and expenditure on a project and for the purpose of project cost management and control, the actual expenditure of both the employer and the contractor clearly needs to be controlled against the original budget figure – as everyone will appreciate from their own personal financial circumstances.

			BUDGET / ORIGINAL ESTIMATE					ACTUAL COSTS					DIFFERENCE			
WBS	Task Name	Duration	Labour cost	Material cost	Plant cost	Other costs	TOTAL COST	Labour cost	Material cost	Plant cost	Other costs	TOTAL COST	Labour cost	Material cost	Plant cost	Other costs
1	Demolition	6 days														
1.1	sub-task 1															
1.2	sub-task 2															
2	Site Preparation	6 days														
2.1	sub-task 1															
2.2	sub-task 2															
3	RC piles	20 days														
3.1	sub-task 1															
3.2	sub-task 2															
4	Drainage works	20 days														
4.1	sub-task 1															
4.2	sub-task 2															
5	Excavation & Support	30 days														
5.1	sub-task 1															
5.2	sub-task 2															
6	Raft Foundation	6 days														
6.1	sub-task 1															
6.2	sub-task 2															
7	Steel frame	21 days														
7.1	sub-task 1															
7.2	sub-task 2															
8	Roof covering	6 days														
8.1	sub-task 1															
8.2	sub-task 2															
9	External wall cladding	20 days														
9.1	sub-task 1															
9.2	sub-task 2															
10	Internal walls & ceilings	25 days														
10.1	sub-task 1															
10.2	sub-task 2															
11	Power supply system	30 days														
11.1	sub-task 1															
11.2	sub-task 2															
12	Lighting system	15 days														
12.1	sub-task 1															
12.2	sub-task 2															
13	HVAC	15 days														
13.1	sub-task 1															
13.2	sub-task 2															
14	IT network	10 days														
14.1	sub-task 1															
14.2	sub-task 2															
15	Internal finishings	15 days														
15.1	sub-task 1															
15.2	sub-task 2															
16	External Works	21 days														
16.1	sub-task 1															
16.2	sub-task 2															
17	Final cleaning & Handover	3 days														

Figure 5.5 Contractor's cost budget.

WBS	Task Name	Duration	BUDGET / ORIGINAL ESTIMATE					ACTUAL COSTS (COSTS COMMITTED)					ESTMATED TOTAL COSTS TO COMPLETE					FUTURE COST EXPOSURE				
			Labour cost	Material cost	Plant cost	Other costs	TOTAL COST	Labour cost	Material cost	Plant cost	Other costs	TOTAL COST	Labour cost	Material cost	Plant cost	Other costs	TOTAL COST	Labour cost	Material cost	Plant cost	Other costs	TOTAL COST
1	Demolition	6 days																				
1.1	sub-task 1																					
1.2	sub-task 2																					
2	Site Preparation	6 days																				
2.1	sub-task 1																					
2.2	sub-task 2																					
3	RC piles	20 days																				
3.1	sub-task 1																					
3.2	sub-task 2																					
4	Drainage works	20 days																				
4.1	sub-task 1																					
4.2	sub-task 2																					
5	Excavation & Support	30 days																				
5.1	sub-task 1																					
5.2	sub-task 2																					
6	Raft Foundation	6 days																				
6.1	sub-task 1																					
6.2	sub-task 2																					
7	Steel frame	21 days																				
7.1	sub-task 1																					
7.2	sub-task 2																					
8	Roof covering	6 days																				
8.1	sub-task 1																					
8.2	sub-task 2																					
9	External wall cladding	20 days																				
9.1	sub-task 1																					
9.2	sub-task 2																					
10	Internal walls & ceilings	25 days																				
10.1	sub-task 1																					
10.2	sub-task 2																					
11	Power supply system	30 days																				
11.1	sub-task 1																					
11.2	sub-task 2																					
12	Lighting system	15 days																				
12.1	sub-task 1																					
12.2	sub-task 2																					
13	HVAC	15 days																				
13.1	sub-task 1																					
13.2	sub-task 2																					
14	IT network	10 days																				
14.1	sub-task 1																					
14.2	sub-task 2																					
15	Internal finishings	15 days																				
15.1	sub-task 1																					
15.2	sub-task 2																					
16	External Works	21 days																				
16.1	sub-task 1																					
16.2	sub-task 2																					
17	Final cleaning & Handover	3 days																				

Figure 5.6 Contractor's cost exposure.

Additionally, good cost management must also focus on future revenues and costs, especially where anticipated technical issues will have a cost effect. For this purpose, traditional financial accounting schemes may not be fully capable of reflecting the dynamic nature of a construction project. Accounts typically focus on recording past costs and expenditures associated with activities and these past expenditures are called 'sunk' costs that cannot be altered in the future and indeed may or may not be relevant as the project goes forward. Since financial accounts are historical in nature, some means of forecasting or projecting the future course of a project is essential for effective management control.

An example of forecasting used to assess the project status is shown in figure 5.4 above, but even this only reflects historical costs and does not adequately show how much more is going to be expended in the future to complete the works. For this, we need to establish additional information as shown in figure 5.6.

In this example, costs are reported in four main categories for each of the WBS elements (or in however much detail is considered necessary), with all the various cost headings for each category:

- *Budgeted cost / original estimate*: derived from the detailed cost estimate prepared at the start of the project.
- *Actual costs / costs committed*: the actual cost incurred to date for each of the WBS elements, which can be obtained from the financial records.
- *Estimated total cost to complete*: the estimated / forecast total cost in each category will be the current best estimate of costs based on progress and any variations, instructions, likely variations etc. since the project commenced. Estimated total costs will be the sum of cost to date, commitments and exposure.
- *Future cost exposure*: estimated cost to completion in each category minus the actual costs paid to date. Commitments may represent material orders or subcontracts for which firm commitments have been made.

For effective project control, commercial managers would focus particular attention on items which indicate a substantial deviation from the original budget (the 80/20 rule). A next step would be to look in greater detail at the various components of these categories to see how they can be mitigated as overruns in cost may be due to lower than expected productivity, higher than expected wage rates, higher than expected material costs, or other factors. If a more detailed assessment is necessary, low productivity may be caused by factors such as inadequate training, lack of required resources, or excessive re-work to correct quality problems. By analysing and evaluating in this way, a strategy for improving cost control can be made – and hopefully built into the company's systems so that it is less likely to happen again.

5.8 Summary and tutorial questions

5.8.1 Summary

Paying the contractor is one of the employer's primary obligations and there are three main ways that this can be done, as explained in section 5.1. Historically, the employer would have paid the Interim Payment Certificates issued by the PM / engineer, but nowadays, in virtually all jurisdictions, the contractor will be informed of the approved amount and be required to submit a formal invoice for payment. This has benefits for both parties.

The build-up of the amounts to be paid in Interim Payment Certificates, as well as the final account, will be developed from the pricing documents incorporated into the project's contract documents. The majority of projects use Bills of Quantities in one form or another, although Bills of Approximate Quantities, Schedules of Rates and Activity Schedules may also be used as pricing documents. Each format has different advantages and disadvantages which should be carefully considered depending on the project circumstances.

When the gross valuation has been calculated by the contract administrator, a retention percentage is normally deducted plus any scheduled repayment of the advance payment to the contractor. If the contractor is in culpable delay and has exceeded the required date for completion, liquidated damages may also be payable and will be deducted by the employer via this certificate.

As the project progresses and payments are made by the employer to the contractor, as well as payments from the contractor to all sub-contractors and suppliers, each party will wish to monitor these payments against their original budgets, as part of their corporate cost control procedures. The types of costs that the employer will incur in realising the development are quite different from the costs that the contractor will incur and both are discussed in section 5.7. Hopefully, on completion of the projects, both parties will be satisfied with the outcome.

5.8.2 Tutorial questions

1. Discuss the advantages and disadvantages of Bills of Quantities produced in trade sections (as per NRM) when calculating interim payments to a contractor.
2. Discuss the advantages and disadvantages of Priced Activity Schedules when calculating interim payments to a contractor.
3. What is the main difference between a time related charge and method related charge? How would these be covered in an Interim Payment Certificate?
4. How can an employer cover the risk of making an advance payment to the contractor?
5. Outline the major differences between a 'Pay Less Notice' and retention.
6. Discuss the major issues surrounding payment for materials on site and off site.
7. Under what circumstances may an employer issue a 'Pay Less Notice'? Does the contractor have any recourse if the notice is not issued properly?
8. How would variations and changes be included in contractor's payments?

6 Variations and changes

6.1 Introduction

As noted in Chapter 1, change and variation during a construction project have always occurred, and this state of affairs is highly likely to continue long into the foreseeable future. The need to make changes occurs for many reasons; in some cases they are unavoidable, because of, for example, unforeseen ground conditions, severely inclement weather, civil commotion or other reasons beyond the control of either party (termed *force majeure*). In other cases the changes may have been generated by the employer who may decide to add something in or take something out of the contractor's scope of work. As well as directly adding to the contractor's scope of work, these changes may also be a cause of delay or disruption to the contractor's programme with the result that extensions of time may be required and additional costs may have to be paid to the contractor – which could be considered as a variation in their own right, or incorporated into the original variation. However, variations do not necessarily cause delay or disruption to the project, so it would be useful to firstly establish the definition of a variation.

Variations generally fall under two headings:

> *Variations to the contract* – i.e. a change to the terms and conditions of the agreement between the employer and contractor, which will require an agreement by the authorised signatories of both parties before the variation can be effective. An example would be if the employer required the defects correction period to be 2 years instead of the 1 year stated in the contract. Clearly, in that particular case, the contractor would want something in return before agreeing to a more onerous obligation.
>
> *Variations to the works* – i.e. changes to the contractor's scope of work from that which formed the basis of the original contract. In other words, a change or amendment to the physical scope of work to be executed. This is usually covered by a term in the contract conditions which allows the employer or PM / engineer to instruct changes.

The former term is often misused as it is sometimes used to refer to variations to the works. As the latter is covered by a specific clause in the original contract, the contractor's agreement is not required before a variation is ordered by the PM / engineer. If there is no variation clause in the original contract, then changing the scope of works would technically be a breach of contract (as construction contracts are 'entire' contracts as mentioned previously), therefore it is in the interests of both parties to include such a clause – for the employer because they can make any required amendments, additions, alterations etc. and for the contractor because they can receive extra payments or additional time to complete the works if necessary.

6.1.1 Definition of variations

In most construction projects, and certainly those which are covered by a standard form of contract, the conditions of contract will contain a definition of variations and the employer's power to omit work applies only to genuine omissions – the employer is not permitted to omit the work in order to have it done by others as this would be both a breach of contract and a breach of good faith (which is normally implied into commercial contracts if not expressly included). The original contractor would then be entitled to claim for the loss of profits that they would have made had they carried out the work themselves. This goes to a critical point in the relationship between an employer and a contractor – generally speaking, if an employer asks a contractor to build a building and goes to the trouble of giving the contractor a full scope of works in the contract documents, the contractor will work out all the resources required to build the building, how long it will take and, therefore, how much it will cost. The unilateral omission by an employer of a large chunk of that work and giving it to somebody else will considerably affect all those calculations and therefore clearly disrupt their scheduling as well as forcing them to lose profit. It is therefore not unreasonable for the contractor to be recompensed for this disruption as well as the loss of profit on the work itself. Fortunately, courts in most legal jurisdictions have agreed with this argument.

6.2 Types of variations

As stated above, most conditions of contract give the employer or PM / engineer the right to order variations or changes to the works. Without this express provision giving the employer the right to order variations, the contractor is not obliged to make any variation and they may insist on a completely new agreement before complying with any request to change the works (entire contract as above) – the employer invites offers for a scope of works, the tendering contractors offer a lump sum price and the employer accepts one of them. Thus the contract is made and any variation to that scope of works would be a contractual breach thus nullifying the entire contract. Clearly, this is unworkable in practical terms, so a contract term is inserted allowing variations or changes to be made.

There are two basic types of variation to the scope of work:

- variations that arise for technical reasons – this type would occur, for example, where unforeseen ground conditions dictate that the sub-structural works would have to be redesigned;
- variations that the employer or their designer desire, but which are not absolutely necessary.

In the former case, the employer would be in some difficulty without the contractual power to order variations which are absolutely necessary to enable completion of the works as the contractor would be able to escape from the original contract price and renegotiate the contract terms. In the latter case, the employer would be unable to take advantage of any desirable changes that may be to their advantage such as reduction in costs or amending the specification to benefit from a value engineering exercise or any new technology or products which have become available.

However, even if there is an extensive variation clause in the contract, the employer cannot make unlimited changes or issue Variation Orders willy-nilly. For example, the employer would not be able to change the works from the construction of a school to the construction of a hotel (although in the present economic climate, most contractors would happily enter into separate negotiations). That is clearly an extreme example, so several of the standard forms of construction contract give guidelines to help in drawing the line between what changes are permitted under the contract and changes that may go beyond the authority of the variation clause.

Some examples are:

1 The contractor may be entitled to refuse to carry out work which exceeds a certain percentage of the contract price (normally 15 per cent). The contractor is usually required to give notice if they believe that the 15 per cent limit is likely to be exceeded, although if they do carry out the excess work, they will almost certainly be able to substantiate a claim for revised rates for the varied work. Clearly each case must be considered on its own merits.
2 If the work is postponed or delayed from, say, summer to winter, the contractor may also be able to renegotiate the rates due to any variable level of productivity in the different seasons, or if they are really lucky, to obtain payment on a daywork, reimbursable cost plus or *quantum meruit* basis.
3 If the work is substantially varied so that the work as constructed is sufficiently different from the original design, the contractor may be able to renegotiate the rates based on a different Method Statement and programme.

As most terms and conditions of contract are written (or amended) by the employer before being offered to the contractor, it is highly likely that, in most contracts, the variation clause will cover quite large variations, so contractors should be aware that too much fine detailing in terms of scheduling the works at the beginning of a project may come to little if large variations are heaped upon them during the construction stage.

6.3 Causes of variations

Variations are caused by a multitude of factors generally related to actions or inactions of one of the parties (the employer, consultants or contractors), or, as mentioned above, due to *force majeure*. The CEBE study in 2004 identified 53 causes of variations grouped under four categories – i.e. owner related variations, consultant related variations, contractor related variations and other variations. See table 6.1 for details of these categories.

Most of the issues contained in this table could be addressed by effective planning and control during both the design stage and the construction stage of the project.

Table 6.1 Causes of Variation Orders in construction contracts (from CEBE Working Paper No. 10)

CAUSES OF VARIATION ORDERS			
Employer related variations	*Consultant related variations*	*Contractor related variations*	*Other variations*
• Change of plans or scope by employer. • Change of schedule by employer. • Employer's financial difficulties. • Inadequate project objectives. • Replacement of materials/ procedures. • Impediment in prompt decision processes. • Obstinate nature of employer. • Change in specs by employer.	• Change in design by consultants. • Errors & Omissions in design. • Conflicts between contract documents. • Inadequate scope of work for contractor. • Technology change. • Value engineering. • Lack of coordination. • Design complexity. • Inadequate working drawing details. • Inadequate shop drawing details. • Consultant's lack of judgement & experience. • Consultant's lack of knowledge of materials & equipment. • Honest wrong belief of consultant. • Consultant's lack of required data. • Obstinate nature of consultant. • Ambiguous design details. • Design discrepancies. • Non-compliant design (with govt regs). • Non-compliant design (with employer req'ments). • Change in specs by consultant.	• Lack of contractor involvement in design. • Unavailability of equipment. • Unavailability of skills. • Contractor's financial difficulties. • Contractor's desired profitability. • Differing site conditions. • Defective workmanship. • Unfamiliarity with local conditions. • Lack of specialised construction manager. • Fast track construction. • Poor procurement process. • Lack of communication. • Contractor's lack of judgement & experience. • Long lead procurement. • Honest wrong belief of contractor. • Complex design & technology. • Lack of strategic planning. • Contractor's lack of required data. • Contractor's obstinate nature.	• Weather conditions. • Safety considerations. • Change in government regulations. • Change in economic conditions. • Socio-cultural factors. • Unforeseen problems.

This clearly proves the old adage that time spent in preparation is never wasted, or the more soporific phrase – fail to prepare, prepare to fail.

Whilst it is virtually impossible to eliminate the incidence of variations in a project, they should be proactively controlled and minimised at the earliest possible time as the potential effects of Variation Orders are not just to increase the costs and time taken for the project, but they can also contribute to the following:

- disrupting progress without any delay
- increase in project cost
- need to hire new professionals to deal with increased workload
- increase in overhead expenses
- delay in payment to the contractor
- quality degradation if re-work is required.
- productivity degradation due to disruption of workflow
- poor safety conditions
- completion schedule delay
- procurement delay of subcontractors
- rework and demolition
- logistic delay
- tarnishing of the firm's reputation
- poor professional relations
- additional payment for contractor
- disputes among professionals.

6.4 Controlling variations

How, then, should variations be controlled? The answer really starts in the design stage and continues throughout the construction stage. The following techniques will help mitigate and hopefully reduce the need for variations.

6.4.1 Variation control in the design stage

Review of contract documents

Contract documents are the main source of information for any project. Therefore, when they are being developed during the latter part of the design stage, it is vital that they are fully coordinated and contain minimum conflicts or ambiguities.

Freezing the design

Any variations in design / scope of works will always adversely affect a project if generated during the construction stage (i.e. post-contract) as the Method Statement and project programme will be disrupted. Therefore, freezing the design is a strong control method, providing of course that the employer is not tempted to unfreeze at a later date. This also requires that the design is complete and fit for purpose at the date of freezing, which unfortunately is not always the case.

Value engineering at conceptual phase

During the design phase, value engineering workshops can be a major cost saving exercise as, at this stage, the design has not yet been confirmed; therefore there is nothing to vary. Value engineering at the concept design stage can assist in clarifying the employer's project objectives and reducing design discrepancies between the various elements.

Early involvement of the contractor in design

Involving people with construction skills during the design stage assists in developing the designs by accommodating different and practical ideas as well as improving the buildability of the design. Issuing Variation Orders during the construction phase is a costly activity and may initiate numerous changes to construction activities which may require numerous changes to planned activities during the construction period. An experienced contractor may be able to prevent such changes by being involved in the design stage.

Employer's involvement at planning and design phase

Involvement of the employer at the design phase would assist in clarifying and embedding the project objectives and ensuring that the project scope of works achieves those objectives. Therefore, this may help in eliminating variations during the construction stage where the impact of the variations is greater.

Thorough detailing of design

A clearer and more transparent design will be understood more readily by all the project participants and also assists in identifying errors and omissions at an early stage. Thorough detailing of design also eliminates the need for variations arising from ambiguities and errors in the design.

Clear and thorough project brief

A clear and thorough project brief helps in clarifying the project objectives to all the participants and reduces the design errors and any non-compliance with the employer's requirements.

Reducing contingency sum

The provision of a large contingency sum invariably affects the participants' working approaches as the size of contingency is a measure of the design completion. Therefore a higher contingency element sends a message that the details are not fully complete and any gaps need to be covered financially. A lower contingency sum shows a degree of confidence that the scope of works covers everything that needs to be done.

6.4.2 Variation control in the construction stage

Clarity of Variation Order procedures

The procedures used for generating, authorising and issuing variations should be agreed and communicated to all parties at the outset of the project and should be clear and transparent.

Written approvals

All variations in the work that involve a change to the original price must be approved in writing by the relevant authority before a Variation Order can be executed by the contractor. This may seem self-evident but in the hectic environment of a construction project, verbal instructions can be forgotten, leaving the contractor without any formal basis to be compensated for the extra work carried out.

Variation Order scope

A Variation Order must – by definition – vary or change something, and that something is the original scope of works. Therefore the VO must contain two separate sections:

a *Omissions* – what is to be omitted from the original scope?
b *Additions* – what is this scope to be replaced by, or added to it?

This would then distinguish between a variation of scope (i.e. items in both (a) and (b)) and a variation due to an increase in scope or further development of the design (i.e. items only in (b) – additions). This would also make it easier to calculate a net cost to the Variation Order and whether the total project cost would end up being increased or decreased.

Variation logic and justification

The reasons for the variation should also be given in the VO – i.e. whether the variation was required or elective. Required changes are those required to meet the project's original objectives whilst elective changes are additional features that are seen to enhance the project, but are not strictly necessary in a value engineering context. Knowing the logic and justification behind any proposed variations assists in giving the parties a sense of shared ownership in the project.

Employer's involvement during construction phase

Involvement of the employer during the construction phase clearly assists in identifying non-compliance with the requirements and possibly in ensuring that the variations are approved promptly. Additionally, the active involvement of the employer during the construction phase would keep them aware of ongoing activities and assist in prompt decision making. Of course, in doing this, the employer must not interfere with the contractual duties of the consultants or impede the progress of the contractor.

Use of project scheduling/management techniques

To effectively manage a variation means being able to anticipate its effects and to control, or at least monitor, the associated cost and schedule impact. Before formal issue of the VO, the cost and time effects should have been thoroughly assessed to ensure that the project budget and schedule are still achievable and that there are no unintended consequences, such as changing the critical path through the project. If there are, these must be managed to mitigate their effects.

Comprehensive documentation of Variation Orders

The timely notification and documentation of VOs (see section on document control in Chapter 1), will ensure that the project not only progresses as efficiently as possible, but also minimises the risk of contractual claims for 'late instructions'. On far too many projects, there is a considerable length of time between when a proposed contract modification is first considered and when the matter is finally issued as a formal Variation Order. The documentation of Variation Orders should be a continuing process throughout the project and would form part of the calculation of interim valuations, regular financial statements of anticipated final cost and, eventually, the final account.

6.4.3 Variation control where design overlaps construction

Prompt approval procedures

In the situation where the design has not been fully completed before construction starts, there is an even greater urgency for prompt procedures in order to reduce the possibility of re-work. This is always a danger in design–build projects where 'later' design decisions can depend on 'earlier' decisions. For example, an alternative specification of doors may mean that the opening size needs to be amended, and the openings could already have been formed. This is particularly important in the area of MEP installations where the 'builder's work', i.e. holes for cables, ducts etc. are positioned on the basis of the design at the time. Subsequent modifications to the MEP design (which always happens) can have a major effect on the builder's work. Because of the fragmented nature of builder's work, any re-work is invariably carried out on a day-work basis, thus increasing the costs even further.

Ability to negotiate variation

The ability of all parties to negotiate variations is an important factor for their effective control. This means that the employer, consultants and contractor(s) can play a part in negotiating variations, which is one of the principles of value engineering – see also Chapter 9.

Valuation of indirect and consequential effects

Consequential effects of early variations can (and do) occur later in the project. Therefore, it is essential to acknowledge this possibility and establish the mechanism to evaluate the consequences of a VO. See the point above regarding door openings and builder's work.

Team effort by all parties to control Variation Orders

Construction projects are multi-party organisations and, naïve as it may sound, the parties should have the objectives of the project as their main priority. Therefore, adverse effects on the project such as unnecessary variations can often be managed at an early stage with appropriate cooperation, coordination and communication.

Utilise work breakdown structure

A work breakdown structure (WBS) is a management tool for identifying and defining work which needs to be carried out. The contractor should be using a WBS in managing their resources, especially on large projects. If a variation involves work not previously included in the WBS, it should be added by the project planner and its relationship with the other WBS elements checked for any conflict or ambiguities. Ripple effects on subsequent activities (i.e. consequential effects mentioned above) can also be traced by the use of a WBS.

Robust variation procedures written into the contract

Most standard forms of contract include clear procedures for the generation, assessment and issue of Variation Orders, which also forms part of the project management system (PMS) of the larger employers and contractors. Most of these procedures include the fair allocation of risk and a fair method of assessing the cost of the variation. These should be written into the Project Execution Plan (PEP – see Chapter 1, section 1.3) which is often a mandatory requirement on larger projects.

Comprehensive site investigation

Below ground conditions contain the majority of uncertainty in a construction project. Therefore a comprehensive site investigation will assist in the proper planning of construction activities as well as accurate choice of foundation type. Differing site conditions are an important cause of delays in large building projects, so a comprehensive site investigation would also help in reducing potential variations in a project.

6.5 Issuing variations

As mentioned previously, Variation Orders / Change Orders should not be issued willy-nilly by the employer or PM / engineer. This haphazard approach sends a clear message that the project is not fully under control, with the risk that cost and time targets will not be met.

Issuing changes or amendments to the project during the delivery (construction) stage should therefore follow a robust procedure to identify, analyse, evaluate and implement the change in a systematic and orderly way. This ensures that only the changes which are absolutely necessary to maintain or improve the project functionality are issued.

Figure 6.1 gives a generalised process model of issuing variation or Change Orders, and a bespoke version of this model should be written into the PEP with a company-specific version included in the Project Management System (PMS).

Change identification

Why does the scope of works need to be changed? What has happened to force this change on the project? As mentioned above, this can either be for reasons beyond either party's control (*force majeure*), a change of mind by the employer, or a consequence of new products / value engineering workshop. Whatever, the reason, the source or driver of the change needs to be identified as this can affect the subsequent process.

Change request issued

Normally such change requests would come from the contractor and be issued to the employer or PM / engineer if they are the formal authority for VOs. However, it is

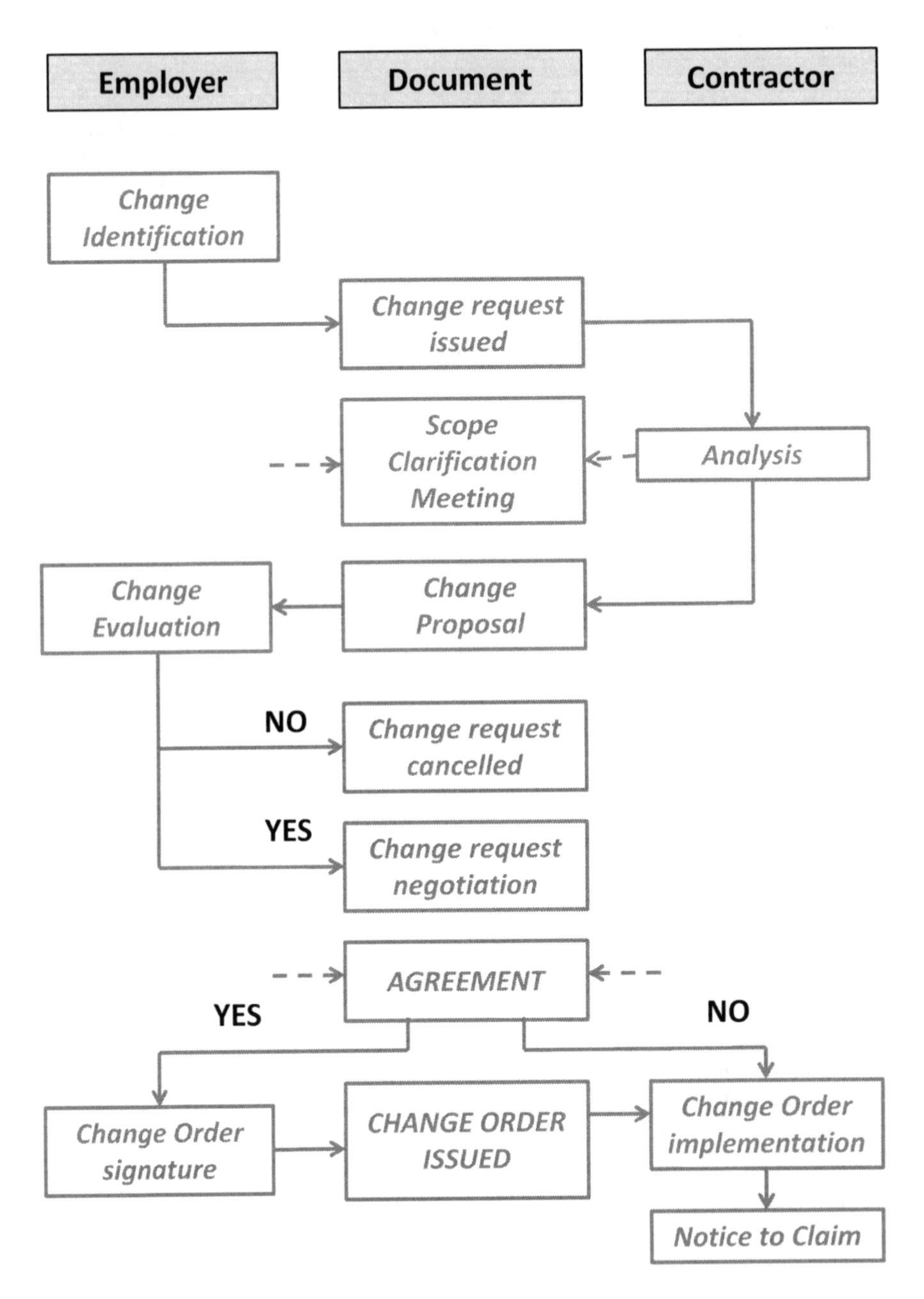

Figure 6.1 Procedure for issuing Variation / Change Orders.

also good practice for either party to inform the other that a change / variation is in the offing – in fact, this 'early warning' is written in to the NEC / ECC contract for all compensation events, as this form of contract requires that both parties work in a spirit of mutual cooperation.

Change analysis

The proposed change should be analysed for any subsequent effects on later operations or programming. These effects could be any of the following:

Dimensional – New or replacement products may have different sizes, shapes, weights etc. from those originally envisaged and designed. This may require some redesign of adjacent or supporting members.

Programming – Variations commonly have a major effect on the programming / scheduling of the works and, depending on when the work is scheduled to be carried out, may delay the operation of the work in question. Whether this requires an extension of time (EOT) is another question, since EOTs are given for delays caused by certain relevant events and the project must have been delayed by that event alone (concurrent delays by non-relevant events cause no end of arguments). Delays to operations on the critical path will automatically cause delay to the project, but delays to operations not on the critical path may not delay the project if there is float attached to the operation. Delay analysts and claims consultants make a very nice living out of trying to sort these issues out on complex projects. See Chapters 10 and 13 for further detailed discussions.

Financial – Variations and changes will clearly have a financial effect on the project as there will be omissions and additions to the scope of works. These omissions and additions will be valued by one of the procedures in section 6.6 below and the total project costs amended accordingly. One of the reasons to include a contingency sum in the contract is to make allowance for possible extra costs due to variations, so that the final cost to the employer should not be too much of a shock from the original cost advised to them.

Change proposal

When the change has been fully analysed, a final proposal can be made which should be clearly understood by all parties. Although this is not yet the formal Change Order, this proposal will allow everyone to make due allowance in their project management, so that when the Change Order is formally issued, it merely acts to tie up the paperwork. The change and all its effects should have already been assessed and managed.

Change evaluation

The change proposal should be the final decisions, so this evaluation stage will be a formal notification of the effect on all the parties, so that nobody is left in any doubt.

Change Order issued

Issue and transmission of the formal Change Order / Variation Order, for document control purposes as mentioned above.

Implementation of change

Actually putting the change into practice, which will hopefully be programmed, scheduled and costed as described above.

6.6 Valuing variations

Under most standard forms of contract, there are three main procedures available for the valuation of variations and also for work covered by provisional sums – although provisional sums do not strictly need a Variation Order as, by definition, they are required to be firmed up for the final account anyway. However, in most cases it is administratively convenient to include the instructions for work covered by provisional sums with instructions to vary the works – they are both instructions by the PM / engineer and both also require the contractor to do something they didn't know they were going to do at the beginning. The three procedures are:

6.6.1 By application of valuation rules laid down in the contract

When rules for the valuation of variations are included in the contract, they will normally follow a strict priority:

a Use the rates in the contract BOQ to value the measurement of additions and omissions from the scope of works.
b If there are no identical items in the contract BOQ, 'star rates' may be calculated using the same estimating principles, spread percentages (i.e. overheads and profit) which were used in the contract BOQ.
c Build up a new rate for the item from first principles (labour, materials and plant costs with add-on for overheads and profit).
d Use daywork rates included in the BOQ for labour, materials and plant with the contractor's percentage addition, which should also be stated in the BOQ.

The overriding principle is to maintain the competitive pricing structure which the contractor employed to calculate the rates in the BOQ at tender stage. If this proves impossible, daywork rates can be used as a last resort as this effectively prices the variation on a cost plus / reimbursable basis with a very high add-on percentage which is intended to take account of any disruption caused to the contractor's programme. If variations are managed properly and incorporated into the project programme early enough, there should be little disruption to the contractor's work.

6.6.2 By agreement between employer and contractor

As this is a commercial contract, the parties are free to agree either the price to be paid for variations or, more fundamentally, any alternative rules / methods for their valuation. Provided such agreements are properly made they will override the contractual terms for valuation and any agreement on price will be legally binding on both parties. Such agreements must be carefully made if all financial implications of the variation are to be taken into account, which is not as easy as it first appears, and sometimes agreements are later wholly denied or, more often, the parties later find themselves arguing about precisely what was or was not included in the agreement.

Therefore, it is essential for any such agreements to be recorded and / or confirmed in writing and signed by the authorised signatories.

6.6.3 By reference to an accepted variation quotation

Many of the common standard forms of contract (JCT-SBC, NEC and FIDIC) allow the contractor to give a variation quotation before the PM / engineer issues an instruction. This has the benefit of rolling up all the costs associated with the change – direct costs such as labour, materials, plant and preliminaries, together with all indirect costs associated with the change, such as head office overheads, any delay, disruption and programming requirements etc. Therefore all 'impact costs' of the variation must be included in the quotation. Clearly, any request by the PM / engineer for a quotation must be accompanied by sufficient information to enable a quotation to be given by the contractor. If the contractor considers the information provided insufficient or deficient, they usually have a set time limit from receiving the instruction to ask for further information.

The quotation must be submitted in compliance with any particular requirements of the instruction and must in all instances state, with full particulars and calculations:

- a the full price for the variation including the cost of all necessary associated work and preliminaries;
- b any associated loss and/or expense;
- c a fair and reasonable amount for the costs of preparing the quotation; and
- d any adjustment of time required for completion (extended or reduced time for completion.

If specifically asked to do so the contractor must also give indicative information concerning:

- e additional resources that will be required; and
- f any proposed particular methods for carrying out the variation concerned.

Following receipt of the quotation, the employer or PM / engineer will have a set time limit to decide whether or not to accept the quotation. If it is accepted the PM / engineer will issue a Confirmed Acceptance instruction. Following the normal rules of offer and acceptance, the employer cannot accept a part or parts of the quotation and should accept it all or nothing. However, practically in such situations it is likely that negotiations will take place between the parties. Using the same principles, the contractor can withdraw the offer at any time before it is accepted, but this rule may be tempered by express terms in the main contract that any quotation is held open for a minimum duration.

Interestingly, the JCT contract allows the final date for completion to be brought forward following such a contractor's quotation. Previously, the final date for completion could only be extended or held the same, even if work was omitted from the contract. As the contractor is giving the quotation, they are given the opportunity to reduce the duration for any reduced workload, in order to be able to use their resources as efficiently as possible on other projects.

6.6 Summary and tutorial questions

6.6.1 Summary

Variations and changes are a fact of life in all construction contracts. Therefore in order to maintain management control of the project – i.e. effective construction contract management – any required changes to either the original contract or the original scope of work must be assessed and evaluated before being formally issued to the contractor.

There are many causes of variations and changes required to the scope of works of a contract, as the works will be done in the future, so unforeseen events happen, new products appear on the market and employers can have a fickle habit of changing their minds about what they want to be included in the project. Whatever the cause of the variation, the process of amending the contract documents and incorporating the changes needs to be rigorously controlled in order to maintain control of the overall project. This involves change identification and analysis, change proposal, change evaluation, change instruction and, finally, change implementation. In many ways, this replicates the principles of configuration management used in the IT and process engineering industries.

As most construction standard forms of contract include appropriate rules for variations and changes, there are certain rules laid down for the valuing of the changes, which would either be by the application of rates for the work needed for the variation, or by the contractor giving a quotation which the PM / engineer will accept or not. The NEC contract uses the term 'Compensation Event' and the contractor's quotation is expected to include all direct and overhead costs, plus all other impact costs, such as delay and disruption caused by the variation. This means that the contractor will not be able to issue a separate 'claim' for these delay and disruption costs, which effectively reduces the potential for disputes as it puts the delay costs where they should be – on the variation.

Whatever, the method of valuing the variation, the amounts will be included in the Interim Certificates when the work has been carried out and accepted and also included in any financial statements given to the employer to advise the eventual final outturn costs.

6.6.2 Tutorial questions

1 Discuss the essential difference between variation to the contract and variation to the works.
2 Discuss the ways of reducing the various causes of variations and changes to a construction project.
3 What are the main factors to be considered during the design stage to reduce the possible number of variations in the construction stage?
4 Outline the procedure to be adopted to control variations in the construction stage.
5 Discuss the requirements of each stage in the procedure for issuing Variation Orders.
6 Outline the main methods of valuing variations.
7 Discuss the advantages and disadvantages of including the delay and disruption costs with the direct costs of the variation.
8 What should be included in the 'impact cost' of a variation?

7 Sub-contracting

7.1 Introduction

A main contractor may engage another company in order for that firm to undertake a specific part of the main contractor's scope of works. Whilst this concept of sub-contracting is not new, it has become more prevalent in the modern construction industry, due to the complexity and specialisation of the technology and the greater efficiency of operations that specialised companies can offer the project, thereby hopefully reducing the project costs. There is consequently a wide variety of specialist firms, together with non-specialist but local firms operating as sub-contractors within the construction industry in all parts of the world. A further reason for the use of sub-contractors is the increased flexibility afforded to main contractors who do not need to invest in the specialist plant and equipment needed for every aspect of construction operations. However, this flexibility must be weighed against the need to select, manage and control appropriate sub-contractors; therefore consideration must always be given to both the benefits and risks associated with sub-contracting.

These issues may of course be settled in the contract between the main contractor and the sub-contractor and there are a variety of general principles applicable to sub-contractor relationships. First, the main contractor remains responsible to the employer for all aspects of the sub-contract; so the main contractor is still responsible for time, quality and payment in accordance with the contract between the main contractor and sub-contractor regardless of any issue that could arise between the main contractor and the employer. This will of course depend upon the terms of the contract between the main contractor and a sub-contractor, and may also depend on the terms of the contract between the employer and main-contractor. However, they are nonetheless two separate contracts in law and the matching or integration of 'back-to-back' provisions and obligations can create their own problems and will often prove difficult to manage in practice.

7.1.1 Privity of contract

As can be seen from figure 7.1, there is normally no direct contractual link between the employer and the sub-contractor through the standard contractual supply chain. In a simple contractual relationship (i.e. with no complex terms and conditions), the main contractor does not act as the agent of the employer to any sub-contractors, and conversely the employer's rights and obligations are in respect of the main contractor only.

There is therefore no formal direct legal relationship between the employer and sub-contractor.

The employer therefore cannot take legal action themselves against the sub-contractor if the sub-contractor's work is defective, lacking in quality, or causes delay to the works. On the other hand, the employer is only obliged to pay the main contractor and so sub-contractors cannot take legal action against the employer for the sub-contract price even if the main contractor defaults or becomes insolvent. These concepts arise because of the general principle that there is no privity of contract between the employer and a sub-contractor; however, the employer, main contractor, sub-contractor and supplier relationships are rarely this clean cut in practice.

The employer may wish to influence the choice of a sub-contractor to be employed on the project, or indeed the terms on which a sub-contractor or supplier is engaged. An employer may also wish to use a specific firm or insist on the main contractor choosing from a list of 'approved' firms to which the main contractor may or may not add further names. The main contractor may only be prepared to sub-contract work on particular terms, or may wish to limit the main contractor's risk or payment obligations. In respect of payment, the main contractor may further wish to limit their exposure by part-paying sub-contractors or sharing the risk of the employer's insolvency. The contractor is therefore normally liable to the employer for any default of the sub-contractor. When a contractor engages a sub-contractor they are simply delegating the performance of the works to someone else and this delegation, together with the management and control procedures, must be clearly outlined in the sub-contract documents to avoid doubt and ambiguity during the construction stage.

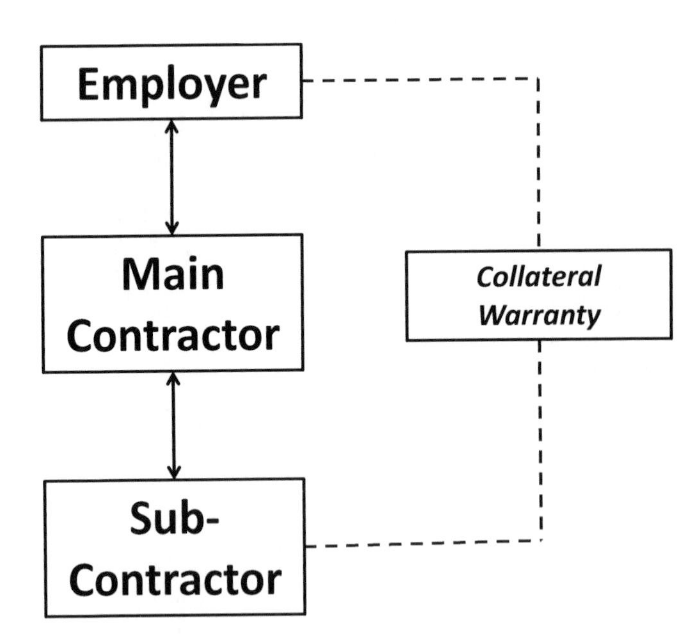

Figure 7.1 Simple contractual relationships between employer, main contractor and sub-contractor.

7.1.2 Personal or vicarious performance

Some contractual obligations are clearly impossible to delegate, and an example could be wishing to have your hair cut by a particular hairdresser, or purchasing a ticket to see a particular singer or entertainer. In these cases, if the particular hair stylist or particular entertainer is not available, then the contract is void, even though the salon is still in business or the concert still takes place. Unfortunately, this is rarely the case in the construction industry and in most cases it will be possible to sub-contract some (or all) of the work so that the contractual obligations are effectively being performed vicariously.

Further issues that arise include the incorporation of the main contract terms into the sub-contract ('back-to-back' provisions as mentioned above), which is often carried out with limited success. Sub-contract terms themselves are often incorporated unwittingly or due to time pressures, which can result in the concept known as the 'battle of the forms' where the terms of a sub-contract or supply of goods and materials are printed on the back of the last form to be sent or received – i.e. the 'offer' which is then 'accepted' by the issue of a Purchase Order. A further issue is the incorporation or otherwise of the dispute resolution procedure, in particular arbitration, which may not have been in the minds of both parties at the outset.

7.2 Relationship between main contractor and sub-contractor

A distinction is often made in the construction industry between 'domestic' sub-contractors and 'nominated' sub-contractors. This distinction is taken to mean that a domestic sub-contractor is totally selected and employed by the main contractor, for whom the main contractor is solely and entirely responsible, whilst a nominated sub-contractor is one selected by the employer but employed by the main contractor. If a sub-contractor is nominated then the employer usually retains some liability and responsibility and this has caused major legal problems in the past, so that most modern standard forms of contract do not include any provisions for employer nomination of sub-contractors. In addition, some standard forms of contract, notably the JCT Intermediate form have created a further category – that of 'named' sub-contractors.

In this case, the employer names one or more preferred companies and the main contractor may add further companies to a sub-contract tender list. The work is then tendered to this overall list and a sub-contractor is selected by the main contractor using their standard selection procedures. The sub-contractor is therefore treated as a domestic sub-contractor thus avoiding the employer liability disadvantages of nomination, but giving the employer some element of involvement in the selection process. The obligations between the sub-contractor and main contractor will be set out in the terms and conditions of the domestic sub-contract and each family of standard forms of contract (JCT, NEC, FIDIC etc.) will have standard conditions of sub-contract for use with their main contract conditions.

As stated above, the contract between the sub-contractor and the main contractor is a completely separate legal agreement from the main contract between the employer and main contractor, although standard forms of sub-contract only tend to be used on relatively large projects and where the sub-contractors have the appropriate commercial skills to be able to understand them. Sub-contracts are usually formed by way of an exchange of letters, or more frequently by the main contractor issuing a Purchase Order (PO) or Letter of Award (LoA) to the sub-contractor. This PO or LoA then seeks to incorporate the terms of the contract, often by reference to some standard terms and conditions printed on the back of the PO or attached to the letter. This can be a very tricky position for both the main contractor and sub-contractor and may lead once again to the 'battle of the forms' to decide which terms and conditions were, or should have been, contained in the offer and which in the acceptance. Good work for construction lawyers.

7.2.1 Incorporation of main contract terms

Far more interesting questions arise when the main contractor attempts to incorporate the main contract terms into their sub-contracts – the 'back-to-back' arrangements mentioned above. A total replication of the main contract terms is, of course, likely to cause a wide range of problems, but the main contractor will wish to ensure that there is full coordination in the supply chain for at least the following issues:

- bonds, warranties, guarantees and insurances – to ensure there is full cover and no unnecessary duplication;
- retention percentages – to ensure the sub-contract retention is at least the same percentage as the main contract percentage;
- payments terms – to ensure that the main contractor only pays the sub-contractor after receiving the funds from the employer – this is not the same as a 'pay when paid' clause which is illegal in the UK.

Delay

An issue relating to the incorporation of terms into a sub-contract is about who takes the risk in respect of any project delays. It may sound common sense that a term should be implied into the sub-contract to the effect that the sub-contractor should be able to organise their work in an efficient and profitable manner (similar to the main contract that the contractor should progress with the works 'diligently'). Therefore, if the main contractor is issued with an extension of time for a relevant event in the main contract, then the sub-contractor should also benefit from the extension of time if the delay also affects the progress of their works.

Dispute resolution

As the contracts between the main contractor / sub-contractor and the main contractor / employer are quite separate legal agreements, the dispute resolution procedures incorporated into the contracts need not be the same. For example, the main contract may include an arbitration clause, but a sub-contract might not, as the scope of the sub-contract may not have the technical or commercial complexity requiring such a major procedure. If the sub-contract does include an arbitration clause, the main contractor may seek to pass on the dispute to the client, thus requiring two separate arbitration proceedings and the consequent increased legal costs. Not a happy state of affairs, especially as the different arbitrators may reach different decisions. Fortunately, the main contract and most sub-contracts are likely to fall with the remit of the Construction Act (if located in the UK or based on English law), therefore adjudication will be the first point of call for any dispute resolution. However, in countries and jurisdictions where there is no requirement for statutory ADR, the dispute resolution procedures for sub-contracts should be considered very carefully to ensure they are appropriate to the main contract terms and can be resolved quickly to avoid any unnecessary delays to the project.

Name borrowing

A related problem area in dispute resolution is that of name borrowing. This procedure allows a nominated sub-contractor to commence arbitration proceedings against the employer, by 'borrowing' the main contractor's name, which can cause some difficulties with both privity of contract and whether both main and sub-contracts contain arbitration clauses. There have been some significant legal cases in this area where a sub-contractor sought redress from the employer by borrowing the name of the main contractor to take the legal action under the main contract. The precise legal nature of name borrowing is fraught with legal difficulties and clearly outside the remit of this book.

A licence to be on the site

The access rights to the site for the sub-contractor should be covered within the sub-contract in the same way that the main contractor's access to the site is covered in the main contract. This is to ensure that the main contractor can retain control over the number of companies on site at any given time. In the absence of any express terms, there will be an implied licence that a sub-contractor will be afforded such reasonable access as will enable that sub-contractor to carry out and perform their sub-contract work. This also applies to all welfare, canteen, toilet facilities etc. on site which will be covered by both the main contractor's and sub-contractors' preliminaries.

Withholding payment, Pay Less, set-off and abatement

'Set-off' is the procedure whereby the main contractor is able to deduct money from sub-contractor A in order to pay sub-contractor B for any acts or omissions of sub-contractor A. In general, a main contractor can only do this by the use of very clear wording in the sub-contract, and in the UK, consideration must also be given to the Construction Act if an appropriate Withholding Notice or Pay Less Notice has been given. This requires that a notice be served within a stipulated period before the final date for payment, or in default 7 days before the final date for payment. The notice must state that an amount or amounts are to be withheld and the grounds for withholding the amount as appropriate. See Chapter 5 for further discussion of this issue.

7.2.2 'Pay-when-paid' and 'pay-if-paid' clauses

The further you travel down the contractual chain, then complaints about the ability to receive payment increase proportionately. Payments to sub-contractors is a regular theme in the technical press and not only concerns the sub-contractor's ability to be paid, but also to protect themselves against insolvency of those above them in the contractual chain. Also discussed regularly are such issues as 'pay-when-paid' clauses, 'pay-if-paid' clauses, the ability to recover retention monies from the main contractor and lengthy payment periods (so the main contractor holds on to the funds for longer thus relieving their own cash flow).

Pay-when-paid clauses mean that the sub-contractor will only be paid by the main contractor when the main contractor themselves receive the appropriate payment from the employer. This is now illegal in the UK (the legal term is that the clauses are 'not effective' – meaning they cannot be enforced in court, unless it is due to the employer's insolvency) although they do find their way into sub-contract conditions in other parts of the world. In this case the sub-contractor will have a commercial decision to make – do they accept the clause and run the risk of delayed (or non-) payment, or challenge the clause and risk future relationships? There have been several legal cases around the world that have sought to show that a pay-when-paid clause is actually a pay-if-paid clause, since if the cheque doesn't arrive from the employer, the main contractor clearly hasn't been paid, so cannot pay the sub-contractor (i.e. pay *when* paid). The courts have generally rejected this argument saying that the debt still exists under the sub-contract.

However, a pay-if-paid clause has more certainty in that the main contractor will only pay the sub-contractor *if* they receive payment from the employer. Clearly, there also needs to be some indication of time as well – how long after payment has been received from the employer will they pay the sub-contractor?

In the United States, a similar clause has been construed as merely postponing payment for a reasonable time. The clause does not mean that a sub-contractor is not entitled to payment because of (say) the employer's insolvency.

7.3 Relationship between sub-contractor and employer

In addition to the particular day to day issues arising from sub-contracting, whether as a domestic or nominated sub-contractor, it is worth separately considering the potential liability of a sub-contractor towards the employer. There may be some form of contractual relationship between the two parties – a direct contract, or collateral warranty, or an implied contract, as well as a liability in tort. These issues are now considered below.

7.3.1 Sub-contractors liability to the employer – collateral warranties

A collateral warranty is a direct contract normally between a sub-contractor and the employer – see figure 7.1 above. This procedure is quite common in the construction industry and refers to the warranties (usual executed as deeds) that sub-contractors are frequently required to provide in favour of employers and other third parties such as end-users, tenants and funding institutions. A collateral warranty is normally a formal written document, which is enforceable in court and many of the standard forms of contract (e.g. JCT) include templates for sub-contractor collateral warranties documents.

Alternatively, as it is a contract a collateral warranty may be formed in a more informal way, perhaps in correspondence or even orally, as long as the usual legal requirements for a contract are satisfied. In other words, there must be a clear offer which has been accepted, certainty as to the subject matter and an intention to create legal relations, together with consideration. The requirement for valuable consideration creates the greatest difficulty in this scenario. However, in practice valuable consideration exists where the employer insists upon the main contractor entering into a sub-contract with a particular sub-contractor after the warranty has been given by that sub-contractor to the employer. In that situation the employer will be able to sue the sub-contractor or supplier for any loss caused by breach of the warranty. Clearly, care must be taken with correspondence during the negotiating and tender period for sub-contractors.

Duty to warn

Many contracts include a duty to warn as an express requirement of the contractor (a major obligation under the NEC contract), and even if there is no express term, this duty may still exist in tort. A specialist sub-contractor may have a duty to warn where, for example, the design of their portion of the works is defective, and the specialist nature of the sub-contractor's work is such that they ought to recognise the defect. Any contractor, being an expert in their field, would normally have an implied duty to warn against issues within their specialisation.

7.3.2 Employer's liability to sub-contractors

In the good old days, when sub-contractors could be nominated by the employer, the standard forms of contract provided for direct payment from the employer to the nominated sub-contractor in certain circumstances and this right was only available where it was expressly set out in the contract, i.e. there was no implied right to direct payment. The reason for direct payment provisions was to protect the nominated firm in case the main contractor became insolvent or ceased payments to the sub-contractor for other reasons. The main purpose for this facility was that nominated firms usually dealt with critical parts of the work and the employer may lose much more money if the main contractor becomes insolvent and the work ceases. The replacement of a main contractor is an expensive business and can also cause significant delay to the project. However, the replacement of a specialist nominated sub-contractor can also cause considerable delay and expense to the employer, so in the event of the main contractor's insolvency it will invariably be in the employer's interest to make a direct payment in order to keep nominated sub-contractors working whilst the main contractor is being replaced.

7.3.3 Employer's instructions to sub-contractors

Due to the concept of privity of contract, the employer has no right to issue instructions to sub-contractors to carry out any specific work. Also, the PM / engineer or contract administrator has no authority or power to instruct the sub-contractor, unless this is specifically included in both the main contract and the sub-contract agreement. However, if the employer does instruct the sub-contractor to carry out any works, this could be construed as a separate contract, based upon an express or implied promise to pay the sub-contractor. It is therefore easy to see how complications can arise if this separate contract interferes with the progress of the sub-contractor's scope of work for the main contractor and will lead to all kinds of difficulties, of which a claims conscious main contractor will undoubtedly seek to take advantage. Conclusion – employers should not issue instructions direct to sub-contractors.

A further problem relates to the rule against preferences arising from insolvency legislation. It is not normally possible for an employer to make a direct payment to a sub-contractor as a result of insolvency and at the same time withhold money from the (insolvent) main contractor's account, especially when the payments should be processed through the main contractor, even though these funds may not find their way to the sub-contractor. The problem, therefore, for the employer is that they may end up making the payment twice. Once to the main contractor (to be added to the assets distributed to preferential creditors under the rules of *pari passu*) and then again to the sub-contractor if they want the sub-contractor to finish the work. Generally speaking, most if not all unsecured creditors to insolvent construction companies receive little, if anything, in the way of payments. Otherwise they wouldn't be insolvent.

However, this rule does not apply where there is a direct contract between the employer and sub-contractor (which also means they are not technically a sub-contractor) such as partnering agreements or when the employer has guaranteed payment to the sub-contractor before they agreed to work for the main contractor. Under French law, for instance, sub-contractors are protected against non-payment by the main contractor and can, in certain instances, claim against the employer for amounts due under sub-contracts. The employer would therefore normally require the contractor to obtain a payment guarantee bond in favour of the sub-contractor. The effect of this law is also important for international contracts where the parties choose French law, as this rule will automatically apply. So, a sub-contract being carried out in another country (other than France, but subject to French law) will still benefit from this provision.

7.3.4 Control mechanisms for the employer

A variety of control mechanisms may be used by the employer or PM / engineer in a main contract in order to attempt to control the extent of sub-contracting. These include:

- Prohibition clauses
- Approval procedures.

Prohibition

The main contract may contain a prohibition or limitation on the main contractor's ability to sub-contract. For example, the JCT Standard Building Contract requires the contractor to obtain consent before sub-contracting, in order to provide some level of control by the employer in respect of the portions of the works that are sub-contracted. In addition, it also provides the employer with an opportunity to identify which elements of work are being sub-contracted and to whom.

The PM / engineer (architect / contract administrator in JCT contracts) or employer may wish to withhold consent where they have had some particularly bad experience with a sub-contractor, or given the particular circumstances and nature of the works it is unreasonable or impractical to sub-let a particular part of it. The assignment of a contract, however, is normally prohibited without the written consent of the other party.

Approval procedures – 'or other approved'?

Many contracts still contain the outdated and dangerous provision that materials or components are to be obtained from 'Company X or other approved' although this is now much less common following competition laws and anti-trust regulations in many countries. The term 'or other approved' does not provide the contractor with any additional rights and the employer or its representative is usually within their rights to refuse consent to use materials or components other than those specified. However, on the other side of that coin is that if the specified supplier fails to deliver and causes delay to the project, the contractor may have cause for a claim against the employer. If the project contains a process for 'materials approval' by the supervision consultants then great care must be taken that this consent procedure does not generate delays.

Negligence

In common law jurisdictions, a sub-contractor may have a duty of care to the employer or indeed future occupiers and / or owners of the building in respect of personal injury and property damage to other property for any negligence on their part, although this will usually extend only to physical damage and not economic loss (such as loss of profits) unless the employer or future occupier has a contractual relationship with the sub-contractor via a collateral warranty. Of course, nothing is ever this clear cut and a sub-contractor (especially one with a particularly specialist product or service which they are engaged to carry out) may have a higher duty of care to the employer and the bar to what is considered to be negligence is consequently increased. Given all this, and to be absolutely sure, it is still in the employer's interests to create a contractual relationship with sub-contractors via collateral warranties.

There is a general principle in many legal jurisdictions that a contractor may discharge their duty of care to an employer by delegating it to an independent sub-contractor and this principle would be even stronger if the employer appoints the independent contractor separately or they are a nominated firm. However, it is fairly obvious that the contractor must select the sub-contractor carefully to ensure that the firm has the appropriate capabilities and there may also be some non-delegable duties which the main contractor is not permitted to pass on to others.

7.3.5 Rights of third parties

If there are no collateral warranties, then as we have seen, the doctrine of privity of contract means that a contract cannot confer rights nor impose obligations arising under it on any person except the parties to the contract. The general rule comprises two factors: the first is that a party cannot be subject to an obligation by a contract to which they are not a party, and secondly, a person who is not a party to a contract cannot claim any of the contractual benefits. This principle has been slightly amended now, especially in the UK with the Contracts (Rights of Third Parties) Act 1999, which attempts to give contractual benefit to appropriate third parties (such as tenants, end-users etc.) but only in certain circumstances. This principle has since been taken up by other countries and jurisdictions with similar common law system – most notably Singapore and Hong Kong.

For a third party to be able to sue for losses requires the following points must be proved:

1 that the loss was foreseeable, and the contractor's original breach would cause loss to later owners;
2 that the contract must prevent an assignment;
3 that a third party must have no other cause of action (for example, a collateral warranty) and;
4 that 'substantial damages' had been incurred by and will be for the benefit of the third party subsequent owner.

7.4 Sub-contracting v. assignment

Assignment relates to the total transfer of rights and obligations in a contract from the 'assignor' to another party, the 'assignee'. Virtually all standard forms of contract specifically prevent the contractor from assigning the contract to another party. For example, the FIDIC Red Book prevents either party from assigning the contract without the prior agreement of the other party and this principle is mainly mirrored in the other major standard forms.

What then does 'assignment' actually mean and what is the difference compared with merely sub-contracting? As contracts legally are an exchange of value, they must therefore contain both benefits and burdens to both parties. In construction projects, the burden on the contractor is an obligation to complete the work and their benefit is the right to receive payment. The burden on the employer is to pay and the benefit is to receive the completed building or facility. See table 5.1 in Chapter 5

The fundamental principle of assignment is that a benefit or burden is handed over to another party, whilst the contract is still in existence. This of course would require yet another contract – a deed of assignment – and it is that which requires the consent of the other party. So, a contractor cannot assign their ability to complete the works and a debtor cannot be relieved by simply assigning the burden of payment to someone else.

A contractor may however assign the benefit of receiving future income (e.g. retention money) in order to obtain credit from suppliers, or immediate cash from a funding institution, but this does not relieve them of the responsibility to complete the works or remedy any snags during the defects correction period.

7.5 Design by sub-contractor

Where an employer requires a specialist sub-contractor to carry out work, one of the reasons for this is that the sub-contractor will invariably have the skills in designing the particular work as well as its construction or manufacture. Therefore, the sub-contractor will be performing a specialist design function in addition to the actual carrying out of the works on site. In these circumstances, the design work performed by the specialist sub-contractor is the normally subject of a direct warranty from the specialist sub-contractor to the employer or end-user to ensure that it is fit for purpose and achieves the desired functionality. If the carrying out of the work on site is sub-contracted by the main contractor to the specialist firm, the extent to which the main contractor is liable for defects in the workmanship of the sub-contractor will depend on the precise terms of the various contracts. Therefore great care must be taken that specialist firms which take responsibility for both design and construction are clear to whom they owe the warranty.

This warranty also includes a specialist firm's development of design started by others, as sub-contractors are often required to produce 'shop drawings' or 'installation drawings', i.e. to take the original designer's concept design and produce detailed working drawings to enable the components to be manufactured and installed.

Because of the specialist nature of their expertise, it is likely that there will be some design development required from the concept design provided to them. If there is any element of design development, then any further drawings are usually considered part of the original design unless there is a specific requirement on the sub-contractor to contribute to or complete the design. In other words, the original designer retains professional responsibility, unless expressly provided for in the contract.

On a more general level, the usual concepts of satisfactory quality, fitness for purpose, diligently progressing with the works etc. will be implied into a sub-contractor's work unless specifically included as express terms in the contract conditions.

7.6 Summary and tutorial questions

7.6.1 Summary

Almost all construction projects of any reasonable size include sub-contractors to cover specialist or highly technical aspects of the scope of works. The reasons for sub-contracting can range from employing specialist companies with lower unit costs, thus reducing the overall costs to the employer, to employing technical specialist organisations with expertise in, say, HVAC (heating, ventilating and air conditioning) installations. In the latter case, the expert companies will also be able to design the systems or installation based on the employer's requirements.

In terms of contract administration and management, there is no direct contractual relationship between the employer and sub-contractor, therefore all instructions, variations etc. will need to be processed through the main or principal contractor, who also has the responsibility to achieve the overall programme. In order to provide such a contractual relationship, a collateral warranty will be created between the sub-contractor and employer, although these do not affect the contract administration procedures.

Main contractors will clearly wish to replicate the main contract terms and conditions in the sub-contracts, so that any risks and obligations they have accepted will be passed direct to the sub-contractor for that portion of the works. This includes obligations for quality, time and payment.

As far as the employer is concerned, they will want to ensure that, whoever does the actual work, they or the PM / engineer has the same level of control as well as recompense if the work is not up to an acceptable standard.

7.6.2 Tutorial questions

1. Discuss the principle of privity of contract in construction projects and the difficulties it can create for an employer.
2. Is it possible to have a truly 'domestic' sub-contractor?
3. What is meant by 'back-to-back' clauses in a sub-contract?
4. What is meant by 'set-off' in terms of sub-contractors?
5. Outline the major differences between 'pay-when-paid' and 'pay-if-paid'.
6. Discuss the functions of a collateral warranty.
7. Outline the main differences between sub-contracting and assignment.
8. Would it be possible for a sub-contractor to have design responsibility with the main contractor having no design responsibility?

8 Achieving best value and cost reductions

Although this book primarily deals with the construction stage, in order to ensure best value and optimum cost for the project the process must start early in the design stage and the workshops carried out during the construction stage cannot properly achieve their objective without the design stage processes having already taken place. For this reason, the design stage processes of value management and value engineering will also be discussed in this chapter, to illustrate how the construction stage processes fit into the overall picture.

8.1 The value management process

Value management is a process made up of a series of techniques used to evaluate the various alternatives available for the design and production of construction projects and can be applied at all project stages. As with most things in life, it has the maximum benefit when applied proactively, although it can also be used reactively to audit an already prepared employer's brief. The value management techniques can be employed on any section of the project but will clearly have the most benefit on the high-value or critical operations (i.e. the 80/20 rule) although applications in other areas should not be ignored, especially if there is a construction logic link to the high-value operations being considered primarily.

Value management is a more complex process than value engineering (see 8.2 below) since it involves the client and key stakeholders in investigating and defining the main strategic and technical areas of the project. This complexity requires a skilled and qualified value management facilitator, who should not be too involved in the day-to-day project management but should clearly have a technical understanding of the project requirements and outcomes.

The first task to be carried out by the facilitator is a detailed study of the project in terms of the employer's (and other stakeholders') requirements. This can be termed a pre-workshop orientation and its purpose is to:

- allow the facilitator to understand the background of the project in order to establish an appropriate workshop agenda;
- assemble relevant data and information;
- determine the full list of project stakeholders.

In order to do this, the facilitator would use techniques such as interviews, questionnaires, document analysis, post-occupancy evaluation of similar projects etc.

At the beginning, the workshop facilitator will probe the team to establish all issues that are relevant to the project at the stage the workshop is being undertaken. These will normally be related to:

organisation	the identification of the employer's business and how the proposed project relates to their business needs;
stakeholders	stakeholder interests in the project from the perspective of any stakeholders attending the workshop;
context	context of the project within the culture, tradition or social considerations of the employer organisation and near neighbours;
location	the physical characteristics of the site;
community	the potential impact of the project on the local community;
finance	the project financing, including sources of funding, budgets and cash flow;
time	overall issues relating to the timing and duration of the project;
legal and contractual	procurement, planning permission etc.;
change management	value management and project management mean there is a 'change' occurring. Any issues relating to this change are best explored at the beginning;
project parameter	all other issues which may impact on the definition of the project including, for example, issues of design, design freeze dates etc.

8.1.1 Function analysis / function diagramming

Function analysis and FAST diagrams are 'summary' techniques where substantial quantities of information generated by the project team is summarised into short statements to succinctly describe what the project is expected to achieve (i.e. its function).

At the strategic briefing stage the functions may be expressed in diagrammatic form as a FAST diagram (an acronym for Function Analysis System Technique). The diagram shows the overall mission statement on the left side and the functional aspirations of the project in levels to the right of this mission statement. A function diagram for a leisure centre can be seen in figure 8.1 below. The major benefit of this type of diagram is that it can show the 'priority of functions' in terms of level 1, level 2, level 3 . . . etc. A function diagram undertaken at the strategic briefing stage will become a reference document for each workshop thereafter as the project progresses.

At the project briefing stage the functional exercise is undertaken in two parts:

a User or process flow diagramming demonstrates the way in which a user of the proposed facility, or product, will flow through or interact with the completed project. This diagramming technique is a powerful way of eliciting the most efficient size and position of functional space and / or position of equipment or plant. Examples of complex flow diagrams requiring expert compilation are: diagrams of patient pathways in a hospital and the process diagram of a food processing plant. A value management audit of such flow diagrams can lead to innovations resulting in significant efficiency improvements;

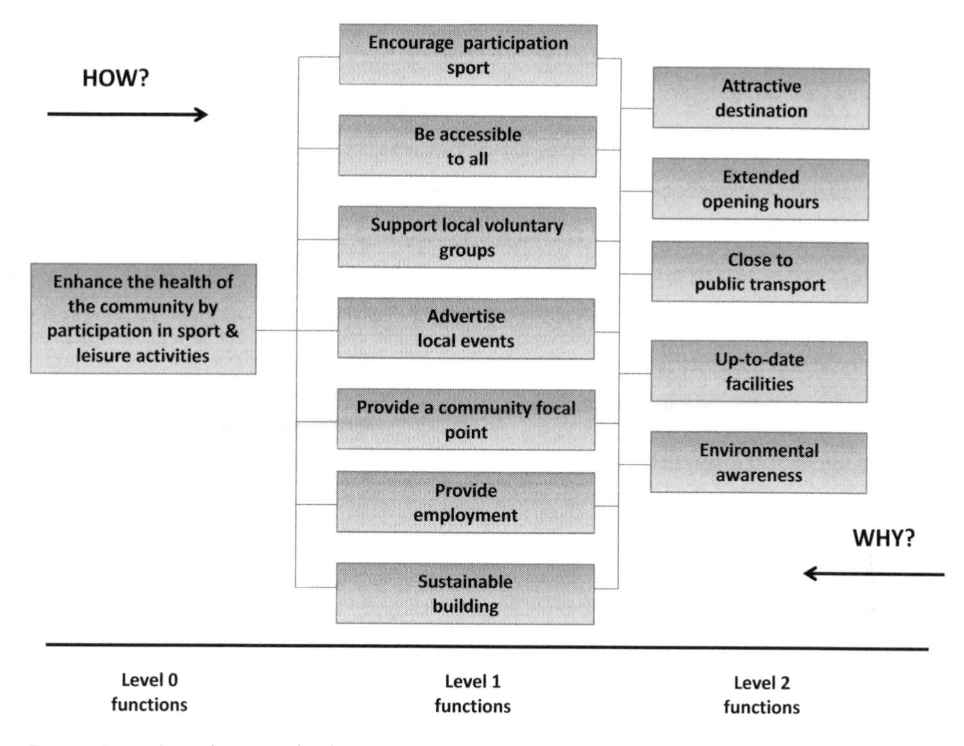

Figure 8.1 FAST diagram for leisure centre.

b Functional space analysis: all activities on the user or process flow diagram will require functional space. The space, normally identified by function, is defined in terms of:

- size specification, for example, a dining room for 12 people;
- quality, normally expressed through a standard of finishing or indicative specification of plant;
- environmental requirement in terms of temperature, lighting, humidity, ventilation/air conditioning, acoustics, etc.;
- IT support.

An effective functional space analysis will simplify the compilation of room data sheets.

Employer's value criteria

There are two client value system techniques in common use which enable the facilitator to explore, with the client representatives of the workshop team, those values held to be important. These values once expressed and made explicit become the criteria against which the project is judged a success. The logic behind the approach is that if the success criterion is known at the beginning then the design can be undertaken and audited at intervals against the criterion, hence improving the probability of a successful outcome. The two techniques are:

- *the time–cost–quality triangle*. This model is discussed more fully in Chapter 1 of *Introduction to Building Procurement* and refers to the three main employer's criteria related to the project in question. Depending on the relative importance of these criteria to the employer, the project would be positioned within the triangle (see figure 8.2).
- *the client value system matrix*. Based on the time, cost, quality principle the value system matrix invites the client representatives of the workshop team to a pairs comparison exercise. The time, cost, quality principle is expanded into nine non-correlated facets to be compared one with another. The facets definitions are:
  - initial capital cost of the project;
  - whole life costing, including operating and maintenance costs of the project;
  - foreign exchange exposure, for international contracts;
  - time: from the present until the completion of the project;
  - local stakeholder opinion including community issues and corporate social responsibility;
  - environment: the extent to which the client wishes to invest in environmental issues beyond that required by law;
  - design flexibility: the extent to which the project is to facilitate changing future requirements;
  - design uniqueness: is this to be a 'statement' project?
  - materials specification: links to whole life costing and environmental issues mentioned above.

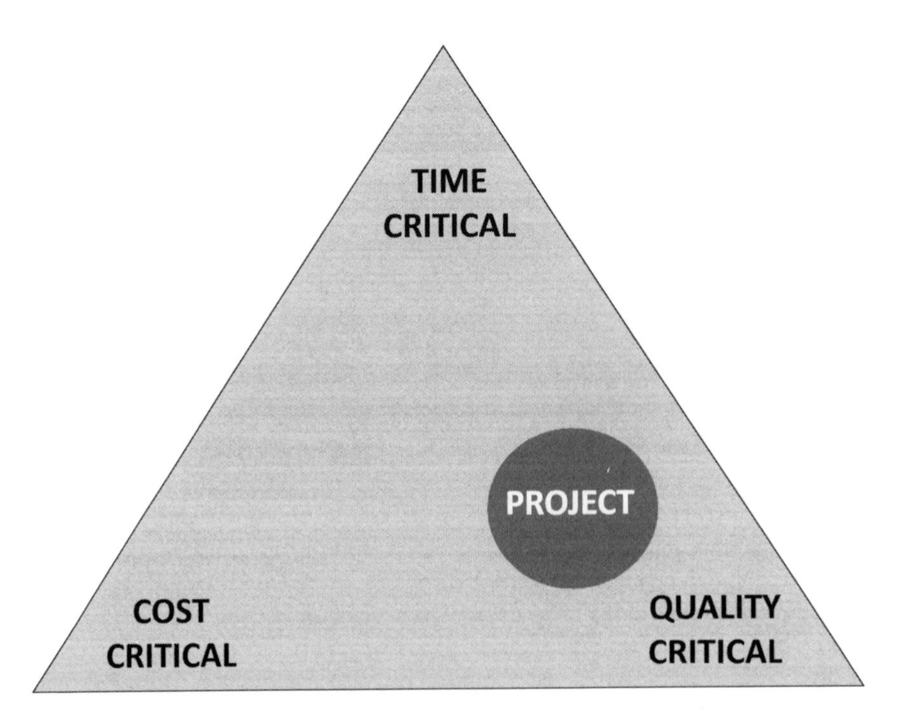

Figure 8.2 Time–cost–quality triangle with project mapped.

The client value system matrix is illustrated in figure 8.3 below. By ranking the facets, the matrix gives an appreciation of those factors which are considered to be most important to the client organisation. The matrix illustrated shows that local stakeholder opinion and environmental impact are the most important criteria with foreign exchange exposure and materials specification being the least important. Clearly, the criteria may change depending on the project. If the project design reflects these issues, then it should be successful.

8.2 Value engineering

Value engineering (VE), as a methodology, was first developed in the manufacturing industry in the USA during the Second World War, when there were severe shortages of materials (and labour) but the factories needed to work at maximum capacity especially if they were involved in the war effort. There was therefore a need for a technique where the input costs could be reduced without reducing the performance (functionality) of the item being manufactured. The term 'value engineering' was adopted, which required the generation of multiple alternatives to the existing solution. As the concept spread across to Europe over the next 20 to 30 years, the term 'value management' was preferred, this being presumably a reflection of the different image and status in Europe between a manager and an engineer. Over time, the term value management came to mean a broader, higher order set of techniques, which also included value engineering.

Value engineering is mainly used in order to eliminate unnecessary costs whilst still maintaining or hopefully improving the function and quality of the design decision

A – Initial capital cost

B	B – Whole life costs (WLC)							
A	B	C – Foreign exchange exposure						
D	D	D	D – Time					
E	E	E	E	E – Local stakeholder opinions				
F	F	F	F	E	F – Environmental impact			
A	G	C	G	G	F	G – Design flexibility		
A	B	H	H	E	F	G	H – Design uniqueness	
J	B	C	D	E	F	G	J	J – Materials specification

A	B	C	D	E	F	G	H	J
3	4	2	3	7	7	5	2	2

Figure 8.3 Value system matrix for a construction project.

or specification and this has developed into an even wider concept nowadays with the concept of 'lean' production, which effectively tries to do the same thing. In construction, this involves considering the availability of materials, construction methods, transportation issues, site conditions, planning and organisation, costs, profit etc. Benefits that can be delivered using VE techniques include a reduction in life cycle costs (costs in use), improvement in quality, reduction of environmental impacts and so on, depending on the employer's priorities.

Although VE should ideally take place at the beginning of the design stage, where the benefits are the greatest and the costs of changing the design minimal, the contractor may also have a significant contribution to make during the construction stage with their skills in programming, scheduling and buildability. Providing that the contractor's proposed changes do not significantly affect the timescales, completion date or project costs, the client would be well advised to seriously consider any contractor's proposal for VE and many contracts actually include provision for the contractor to make such suggested changes in a procedure known as value engineering change proposal (VECP).

The FIDIC Conditions of Contract includes value engineering at clause 13.2, which states:

> The Contractor may, at any time, submit to the Engineer a written proposal which (in the Contractor's opinion) will, if adopted, (i) accelerate completion, (ii) reduce the cost to the Employer of executing, maintaining or operating the Works, (iii) improve the efficiency or value to the Employer of the completed Works, or (iv) otherwise be of benefit to the Employer.

If the contractor's proposal is approved by the engineer, the change will be classed as 'contractor designed' (with the appropriate design liabilities etc.) and will be incorporated as a variation to the contract with any necessary adjustments to the contract sum. If this adjustment has the effect of reducing the contract sum, the contractor will effectively lose out and presumably will only suggest a change proposal to improve the efficiency of their own operations or to score some brownie points from the employer. However, if the contract is on a target cost basis, both FIDIC and NEC state that the target cost will not change, so any reductions in actual cost will benefit the contractor by the percentage gainshare which is included in the contract. There is an expectation in a target contract that contractors will make such proposals and the payments methods are designed to clearly reflect and encourage this practice.

Value engineering therefore involves the following steps, which usually take place in a project value engineering workshop:

- Identify the main elements of a product, service or project.
- Analyse the function of these elements (using FAST procedures – described above).
- Develop alternative solutions for delivering the functions.
- Assess the alternative solutions using pre-agreed criteria.
- Allocate costs to the alternative solutions.
- Decide which of the alternative solutions provides the best functionality per unit cost.

Therefore, VE is an exercise which should involve all parties in the project in selecting the most cost effective solution to a project requirement. It is about taking a wider view (or 'helicopter' view as defined in some management textbooks) and considering the selection of materials, plant, equipment and processes to assess if a more cost effective solution is available that will achieve the same project objectives. The project manager must take a proactive role in both giving direction and leadership in the VE process and must also ensure that time and effort is not wasted – i.e. value engineer the VE process.

Value engineering can therefore be regarded as having two components:

a the provision of the necessary functions of the element being considered at the lowest cost and at the required quality;
b the identification and elimination of any unnecessary costs. This involves the examination of the element to determine those areas which do not efficiently contribute to the required function. In this context unnecessary costs can be defined as cost embodied in the element that:

  - do not contribute to the element's performance;
  - do not contribute to the element's aesthetics or appearance;
  - do not satisfy the employer's or end-user's required features;
  - over provides the employer's or end-user's required features;
  - extends the life of the element beyond that required;
  - does not delight the employer or end-user.

Value engineering construction studies are normally narrowly focused and therefore workshops tend to involve only those few people with the necessary expertise. The workshops therefore are characterised by small teams, are short in duration and are self-contained in terms of decision making, action planning and implementation. The facilitator will normally prepare the workshop agenda as well as a brief report of the workshop outcomes.

8.2.1 Value engineering workshops

Value engineering is normally considered as a series of separate short events throughout the project's life – where particular issues are considered which may improve functional value or reduce costs. These short events (workshops) are undertaken at key stages throughout the project, from early in the design stage, all the way through the end of the construction period, although their effectiveness will necessarily diminish as the project progresses and the works become a physical reality. Each event normally consists of three stages:

- *orientation and diagnostics*: at this stage the aim is to define what the event is trying to achieve and thereby develop the workshop agenda and full engagement of all participants;
- *workshop*: a VE workshop is a formal team activity which attempts to find a 'best value' solution to achieving the functional requirements of the issue under review. Alternative ideas and proposals are encouraged and considered. It is important in this kind of workshop to ensure that all participants feel that they can make a contribution, irrespective of their status in the organisation. The outcomes of this workshop will inform the implementation stage;

- *implementation*: this is where the outcomes of the workshop are investigated and considered with respect to company procedures, cultures etc. Short formal meetings will take place to discuss the conclusions and determine the way forward.

Value engineering workshops follow the following steps:

Identification

The objective of the identification stage is to clearly define the subject under review in terms of a description of the space, element or component. It is common for the appropriate designer, architect, engineer, interior designer etc. to introduce the subject under review and the design philosophy underlying the choice of elements, components and the configuration of space as shown on the drawings.

Function definition

The team having confirmed its understanding of the subject under review, whether that be a space, element or component, the facilitator will ask the team for a list of functions performed by the subject. It should be noted that the same element may have differing functions in different parts of the building. For example, a window at ground level may have a security function whereas a window at the sixth floor level may not. Space, element and component functions are listed and described in a short statement ideally using an active verb and a descriptive noun. For example, a window normally has the function to transmit light and may also have additional functions; control ventilation, permit view, contribute to the aesthetic, secure space etc.

Cost, quality, value

Cost, quality and worth are descriptors attached to the subject under review and should be understood prior to the innovation stage.

Cost represents the monetary whole life cost of the item as included in the whole life cost plan. Often the whole life cost is unavailable and therefore the capital cost from the capital cost plan or tender document is used.

Quality defines the subject under review in the context of client values and design expression. For example, an internal door in a particular location may be defined in terms of its appearance, solidity, fire resistance, environmental attributes etc.

Worth in a VE context is defined as the lowest priced product to meet the functional requirements defined in the function list. Worth therefore tends to ignore certain quality requirements in order to gain an appreciation of the monetary value of the basic utility. For example, the utility functions of a hardwood faced, hardwood lipped flush door could conceivably be undertaken by a simple plywood faced door blank.

The appropriate questions to ask in the workshop are:

- How much is being paid for facets other than utility?
- Is the amount appropriate given the definition of client required quality?

Innovation

With a clear definition of the subject under review and its function, whole life cost (or capital cost), quality specification and worth, the team is now in a position to brainstorm optional solutions to the subject under review. It is important that the brainstorming is allowed to flow without interruption and therefore all consideration or judgement is held until the end of the session.

Cost / benefit appraisal of options

Those optional solutions for the subject under review which, on first impression, appear worthy of consideration are subject to a cost benefit appraisal. This appraisal may take place during the workshop or be undertaken by a team member(s) following the workshop.

Selection and recommendation

If an option appears to give better value for money than the solution incorporated in the existing design then the option is recommended for acceptance within a changed design. The recommendation of an option must always include consideration of the time and cost involved in changing the design and the impact of any delay to the overall project programme.

8.2.2 Construction operations value engineering workshop

Post-contract VE studies are typically short and intensive and attended by those responsible for converting the design into reality in the most efficient manner possible. Issues of assembly are identified by function and resolved by component selection and operational planning. At this stage changes to specified components may be suggested by the contractor by using VECP procedures as described above. The VECP clause may also give the contractor a share in any cost savings through a Painshare / Gainshare procedure which is common in target cost contracts and framework agreements. However, it is important that the rules of VE are consistently applied and decisions do not compromise quality through any reduction in specification.

8.3 Summary and tutorial Questions

8.3.1 Summary

The value management workshop is a formal, team orientated activity, which should be led by an independent, skilled and qualified facilitator, and the independence and skill of the facilitator is a key factor in the successful value management of projects. The facilitator can be an external consultant or a person within the employer's organisation who is suitably qualified and not personally involved with the project under review.

The value engineering workshop, on the other hand, may be facilitated by a knowledgeable and appropriately skilled project manager or cost consultant. In either value management or value engineering, the quality of the facilitation and the most appropriate team are the primary pre-requisites for a successful workshop.

The ACID test is one technique which can be used to select the most appropriate workshop participants:

Authority those members in the employer organisation who have authority to take decisions regarding the project are potentially suitable value management team members. In value engineering studies this authority is more likely to rest with members of the design team.

Consult those members of the client organisation, design teams, specialists and members of stakeholder groups who have the ability to give expert opinion regarding relevant issues should be considered for membership of the workshop.

Inform those to be informed of the outcomes of the workshop should not be included in the workshop team but be included on the circulation list for the Workshop Report.

Do those who are likely to be required to implement the recommendations of the workshop should be included as part of the workshop team. Their inclusion will serve to increase the rate of take-up of innovations generated during the workshop.

8.3.2 Tutorial questions

1. Outline the major differences between 'value management' and 'value engineering'.
2. What are the main purposes of a FAST diagram?
3. Construct a value system matrix for the building you are currently sitting in. Are there any lessons to be learned?
4. What are the main incentives for a contractor to make a value engineering change proposal?
5. At what stage in a project should value engineering workshops take place?
6. Explain the benefits of the ACID test.

Part C

Finalising the construction stage

Completion and close-out

When the contractor has completed all of its obligations, i.e. the works have been built to the satisfaction of the employer (or the PM / engineer), there clearly needs to be a closure to the construction stage, which then heralds the beginning of the occupation / usage stage of the building or facility.

There may be situations where the employer requires to terminate the construction contract before the work has been completed, for example due to the insolvency of the contractor, serious breach of contract, or occasionally 'termination for convenience', and many standard forms of construction contract allow for such early termination. Additionally, situations may also occur when the works need to be suspended for a period of time, for reasons of *force majeure* for example. Although suspension is not technically a contract closure, it is generally included in this area of contract administration.

Suspension and termination before the works have been fully completed is a very messy business and should be avoided if at all possible. There are clearly going to be additional costs for both parties, and many of these costs are unlikely to be recoverable. There is also an effect on the reputation of the party instigating the suspension or termination, as the construction industry is a small world and poor reputations mean higher risks and more difficult working relationships in the future.

However, for the vast majority of projects, the scope of works will be completed by the contractor and in terms of contract management, the major stages of substantial completion, defects correction and final completion will need to be managed in accordance with the rules laid down in the particular contract, before the contract can be finally closed out and everyone moves on.

9 Early termination of contracts and suspension of the works

9.1 Introduction

Termination or suspension of a contract prior to completion of the works is a serious step to take and can have major financial, legal and practical consequences for both parties, so great care should be taken before actually terminating or suspending a construction contract. In very few cases are contracts terminated or suspended by mutual consent, therefore if the contractor objects to an employer's Notice of Termination or Suspension, there could be significant financial effects for the employer if the adjudicator, dispute adjudication/review board (DAB/DRB) or arbitrator later decides that the employer was not in fact entitled to terminate or suspend the contract and acted beyond their powers.

A party's right to terminate or suspend a contract can be given in the contract itself or in the law of the country. The grounds on which a party may do so in law will depend on the governing law of the contract and are often very narrow. In contrast, specific contractual provisions often provide remedies that are greater than or different from those available under the governing law and as the contract is a commercial agreement between two parties, the circumstances can be as wide or narrow as the parties choose to negotiate. I use the word negotiate somewhat advisedly here, since in most traditionally let projects, the contractor is rarely given an opportunity to comment on or negotiate, let alone object, to the conditions of contract for a project.

9.2 Termination by employer

Most standard forms of contract set out a list of circumstances in which the employer would be entitled to terminate the employment of the contractor, together with the procedures that must be followed. These circumstances are normally related to the default of the contractor, for example in abandoning the works, failure to remedy defective work when instructed to do so or failure to provide an adequate performance bond if required to do so. Some standard forms (notably FIDIC) provide for termination by the employer for convenience (i.e. where there has been no default by the contractor and the employer finds that they are not in a position to continue with the project, or possibly just gets fed up with the whole thing).

Clause 15 of FIDIC (both Red and Yellow Books) gives the rules for employer termination and also includes other circumstances where the employer has the right to terminate. These are:

- failure of the works to pass tests on completion;
- failure by the contractor to remedy defects;
- if the contractor sub-contracts the whole of the works or assigns the contract without the employer's permission;
- where the employer invokes *force majeure* or exceptional events beyond the control of the employer;
- where the contractor is released from performance under the law of the country;
- where the contractor becomes bankrupt, goes into receivership, liquidation or other forms of insolvency;
- where the contractor gives or offers bribes, inducements etc.

Interestingly, the first two issues will mainly occur at the end of the project when most of the work will have been done anyway.

A trickier situation is where the contract states that an employer may terminate for 'failure to proceed with the works'. Unless the programme is a contract document, there is a limited amount that the employer can do regarding the contractor's progress of the works except give notice that they are unlikely to achieve the contractual milestones, therefore great care must be taken with this issue and both the employer and PM / engineer would be well advised not to rely on a single instance of failure to proceed, but take an overall view of any general and deliberate slowdown of the works. A particular issue on many international contracts is the – extremely – late payment by some employers, which not surprisingly causes the contractor to slow down the progress of the works due to cash flow difficulties. This often then causes the employer to accuse the contractor of deliberately slowing the progress of the works and a vicious circle is created.

Most standard forms of contract address the issue of bribery and corruption by prohibiting the giving of bribes, gifts, gratuities or commission by the contractor's personnel, agents or sub-contractors, to any person in order to gain an unfair advantage. The employer may also be able to terminate the contract if the actions of sub-contractors, over which the main contractor has no control, amount to bribery or corruption. This facility often excludes the 'lawful inducements and rewards' such as performance related bonuses and profit shares – but only to staff directly involved with the project and these are also normally 'after the event' payments, i.e. payments for performance already done. This is a very topical issue in world trade at present and countries who have signed up to the World Trade Organization agreements and rules are duty bound to reduce corruption and improve governance of commercial contracts. See also discussions in Chapter 16.

9.2.1 Procedure for termination by employer

If the employer terminates the contract 'for cause' (i.e. has a valid contractual or legal reason to terminate), then normally they will be required to give 14 days' notice to the contractor that the contract is being terminated. The contractor must then leave the site, including all off-site areas such as lay-down areas where materials are being stored etc. To show that the employer is serious, it would be advisable to give a further notice to the contractor after the 14 days to the effect that the contract has now been terminated. However, before this course of action takes place, there will hopefully have been a lead-up period when the issues will have been aired between the parties and the contractor has been given an opportunity to take remedial action; if the contract includes early warning procedures, this dialogue will be a mandatory requirement. The fact that the client authorised the project in the first place should mean that they want it finished, so terminating the contractor's employment should necessarily be an action of last resort. However, once the contractor's employment under the contract has been terminated, the employer will need to make arrangements to complete the work by others, so the original contractor will be required to hand over all necessary contract documents, designs etc., leave the site in a safe condition and also allow the assignment of any sub-contracts either to the employer or a different main contractor. Clearly, this will not be a seamless process and sub-contractors in particular may find that their programme is severely interrupted and will invariably seek loss and expense damages accordingly. From whom they seek these damages is an interesting legal argument.

As mentioned above, the contractor must leave the site by the day their contract has been terminated (as they are no longer authorised to access the site, their insurances will also be invalid) as well as handing over any plant, materials, temporary works and equipment to the employer or their representatives. This can be a particularly tricky issue as the plant and equipment may be owned by the contractor and therefore useful on other projects, or else hired from a plant hire company and not owned by them anyway. Some contracts allow the employer to use these items of plant and equipment to complete the project, only handing them back to the original contractor on completion. If this is not the case, any new contractor would need to supply their own plant and equipment in order to complete the project. Depending on the nature of the performance bond (if any), the employer may offset the extra costs of completing the works against the value of this bond, although it is highly likely that the employer would still incur significant extra costs in completing the project by others, which is another reason why they should take great care in terminating a contract. In certain circumstances, the employer may be able to sue the original contractor for these extra costs which are not covered by the performance bond, but that would involve an additional legal procedure, thus extra time and extra non value added costs.

On termination of a contract, a final account needs to be prepared just as it would be when the contract comes to a normal completion. As the contract no longer exists, the employer cannot issue any instructions to the contractor and the employer is additionally responsible for securing and insuring the works until a new contractor is appointed. The final account is a calculation of the works and goods due to the contractor for work executed in accordance with the contract, so any defective work would be excluded and the employer may also withhold further payments to the original contractor until the costs of design, execution, completion and remedying any defects have been established. Clearly, this may take some time.

9.2.2 Employer's termination for convenience (i.e. without cause)

As stated above, the international standard form of contract – FIDIC – gives the employer the right to terminate the contract at any time for convenience, i.e., without any default on the part of the contractor or any other justification. The relevant clause in FIDIC Red and Yellow Book contracts is given below:

> 15.5 Employer's Entitlement to Termination
>
> The Employer shall be entitled to terminate the Contract, at any time for the Employer's convenience, by giving notice of such termination to the Contractor. The termination shall take effect 28 days after the later of the dates on which the Contractor receives this notice or the Employer returns the Performance Security. The Employer shall not terminate the Contract under this Sub-clause in order to execute the Works himself or to arrange for the Works to be executed by another contractor.

So, even though the employer can terminate the contract for their own convenience, they cannot (contractually) do it just to give the work to another contractor. Again, on such termination, the contractor will be required to leave the site and a final account will be calculated.

A termination for convenience clause is generally intended to cover circumstances where the project is no longer required or cannot proceed for whatever reason; for example, if the employer is in financial difficulties and cannot afford to continue to pay for the project. The contractual procedures will aim to restore the contractor to the financial position they would have been in had the project never occurred (hence there is no allowance for profit). It follows that the employer is not entitled to terminate the contract merely to complete the works themselves or by engaging a new (and potentially cheaper) contractor but the contractor cannot claim for profit they would have made had their skills been employed elsewhere although they are entitled to be paid other costs that they have reasonably incurred in the expectation of completing the works, together with the cost of clearing the site and – if required – repatriating staff.

Termination for convenience clauses are invoked by employers quite regularly in the United States and on international projects, but are relatively rare in the UK, where termination for convenience clauses are either not included in construction contracts, or fault-based termination is preferred (as it is generally more financially advantageous). Under the NEC conditions, the very first clause states:

> The Employer, the Contractor, the Project Manager and the Supervisor shall act as stated in this contract and in a spirit of mutual trust and co-operation.

and employers have sought to terminate contracts stating that the contractor has not acted in such a spirit. In common law jurisdictions, there is an implied term that the parties will act in good faith, and terminating for convenience without good cause may be argued to be contrary to the implied term of good faith. Contractors need also to be aware that employers may terminate the contract using these provisions where they themselves are in breach, as a pre-emptive step to prevent the contractor making a claim for breach of contract. This is clearly not what was intended when drafting the clauses.

Given the current global economic climate, it is likely that these clauses will be invoked more frequently as employers find it increasingly difficult to fund projects.

9.3 Suspension by the employer

The instruction to suspend work is normally issued by the PM / engineer, usually at the request of or at least with the consent of the employer. Depending on the actual contract conditions, the instruction to suspend work does not usually have to indicate the justification for suspension, but the PM / engineer will often do so, for record purposes or merely out of courtesy. The use of suspension is not limited to a necessity arising out of specific circumstances – it is a unilateral entitlement of the employer and standard forms of contract do not contain any limitations on how many times the right to suspend the work may be exercised by the employer.

Upon receipt of an instruction to suspend the work, the contractor has an obligation to actually cease performance of the work and consequently will not be paid. Moreover, the contractor has several additional obligations to protect, store and secure the works against any deterioration, loss or damage during the period of suspension.

The contract may or may not provide for a fixed maximum period of suspension; nor is the PM / engineer required to indicate the period of suspension in the notice given to the contractor. However, the contractor is generally given the right to request the PM / engineer's permission to proceed with the work if the suspension lasts longer than 84 days (under the FIDIC conditions). There are various follow-on procedures to ensure that the project does not end up in limbo.

The suspension of work may end at any time by the PM / engineer's issuance of an instruction to proceed with resumption of the work and the contractor is expected to re-mobilise the workforce quickly. If any defects, deterioration or losses have occurred during the period of suspension, the costs will normally be claimable by the contractor, providing they took reasonable steps to avoid the loss. This is also the case for plant and equipment which has been idle during the period of suspension, again providing the contractor has made efforts to re-allocate it. In all cases, the actual contract conditions must be checked to ensure these costs are indeed claimable by the contractor.

During suspension of work, the site will still be regarded as a construction site even if work is not actually being performed. Therefore, the duties of the construction supervision consultant will continue, such as maintaining the construction documentation, preventing unauthorised access and properly securing the site. The contractor remains responsible for the site for the entire time, only that work is not being performed during the suspension period.

As mentioned above, suspension of work clearly involves a gap in the performance of the work and may also give rise to significant costs on the part of the contractor connected with protecting and securing the work during the suspension. Therefore, the contractor may also be entitled to:

- an extension of time for any such delay, if completion is or will be delayed;
- payment of any loss and expense arising from the suspension.

However, the contractor will not be entitled to these rights if the delay or costs are the result of a defect in the contractor's design, performance or materials or in its duty to protect, store or secure the works.

Suspension of construction work is therefore a right of the employer which may prove to be a serious burden for the contractor if a suspension takes place. The costs of suspension will be normally borne by the contractor, up to a period of as long as 112 days, and the contractor will be responsible, at their own cost, for securing the construction site and the existing works and performing their other obligations under the contract, apart from the actual performance of construction work. The contractor will not only not be paid, but may also incur very high costs for a situation which was not of their making and over which they have no control.

When faced with a contract framed in this way, the contractor should at least appropriately secure all agreements with sub-contractors, in order to have an equivalent back-to-back right to suspend sub-contractors' work. Otherwise the costs borne by the main contractor may also include claims by sub-contractors for their inability to access the site to perform their work.

9.4 Termination and suspension by the contractor

Termination of construction contracts by the contractor are extremely rare and on many international projects the provisions included in a standard form of contract for contractor termination (e.g. FIDIC Clause 19) are deleted in their entirety by the employer before contract signature. This doesn't mean that the contractor cannot terminate the contract, but they cannot do so under the contract and therefore will often require a court order, which can be expensive and time consuming.

As stated above, FIDIC Red Book Clause 19 entitles the contractor to suspend work (after giving the employer not less than 21 days' notice) in any of the following circumstances:

- The engineer fails to certify Interim Payment Certificates in accordance with the requirements of the contract.
- The employer fails to provide reasonable evidence of its financial arrangements to pay the contract price, within 28 days of a request by the contractor. Or
- The employer does not pay the contractor in accordance with the contract (for example, the final date for paying the contractor is 56 days after the issue of each Interim Payment Certificate).

Additionally, the contractor may terminate the contract (after giving the employer appropriate notice, unless there has been prolonged suspension or insolvency of the employer) in a number of other circumstances, such as:

- The contractor does not receive reasonable evidence of the employer's financial arrangements within 42 days of suspending for the same reason (see above).
- The engineer fails to issue an Interim Payment Certificate within 56 days of receiving all of the required information under the contract.
- The contractor is not paid an amount properly due in an Interim Payment Certificate within 42 days of the 56 days for payment (i.e. a total of 98 days after the issue of the Interim Payment Certificate).
- The employer substantially fails to perform its obligations under the contract.
- There is a prolonged suspension of the works (typically a period in excess of 84 days).
- The employer becomes insolvent.

Regarding suspension of the works, in both UK and international contracts, the contractor has the right to suspend work due to failure to certify, failure to demonstrate ability to pay or failure to actually pay. Of course, no contractor will terminate a contract lightly, especially if they would quite like more work from this employer, and these provisions only relate to actual contracts which have not deleted them, as mentioned above.

9.4.1 Non-payment by the employer

In many developing countries (and some extremely wealthy countries who are still developing their infrastructure), the issue of non-payment by the employer is a major problem for many contractors.

Suspension of the works

A question that is often asked by them is 'can the contractor suspend works if it has not been paid (under the contract or law)? The answer is a qualified yes, although it will clearly depend on the actual conditions of contract. Under a FIDIC contract the contractor may suspend if payment is not certified or (if certified) payment is not made for 21 days. Notice must be given of the default and then the contractor may suspend or reduce rate of work. The contractor will be entitled to cost and time, and is also under an obligation to resume after payment.

The question of law is often more interesting, as many legal jurisdictions recognise that a contract is an exchange and if one party defaults on their own obligations, the other party can often refrain from performing its obligations until the default is rectified. However, as stated above, this is a high risk strategy as it may not be clear that suspension is intended.

Termination of the works

Similarly, if a contractor has not been paid, does that contractor have an automatic right of termination? Does this change if the contractor has been paid in part? The answer again is a qualified yes, as for instance under FIDIC, if the employer fails to pay sums due to the contractor for more than 42 days after the expiry of the payment periods under the contract. Contractors beware though, as there are conditions precedent before they can go stomping off the site.

The following article has been taken from the technical press in the UK.

S is for suspension, T is for termination of a contract

By Michael Conroy Harris
Building Magazine, 22 July 2011

S is for suspension

However harsh it seems, there is no common law right that allows you to stop performance of a contract if you have not been paid. You can have a contractual right to suspend work if it is clearly set out in your contract and, in the UK, you do have a statutory right to suspend under the Construction Act. You have to give at least seven days'

(continued)

(continued)

notice of your intention to suspend under the Act (but note that this is sometimes extended in the terms of a contract) and your notice needs to set the grounds under which you intend to suspend. Your right to suspend ends when you receive the outstanding payment but you are not liable for the consequences of the suspension. When the changes to the Construction Act take effect on 1 October 2011, you will also be entitled to your costs and expenses of the suspension and any time due to it (such as the additional time taken in remobilising).

While you have an important and powerful right, you have to use this right carefully – make sure that you do not end up being at fault by repudiating the contract. Careful distinction needs to be made between temporarily suspending performance for non-payment and walking away from the contract entirely. In the recent case of Mayhaven Healthcare vs Bothma, a contractor that thought it was properly entitled to suspend for non-payment was held to be in breach of contract when it was established it did not have the right. However, it was not found to have repudiated the contract simply because it mistakenly thought it had a right it did not have. You should not assume that this will always be the outcome, as it will depend on the circumstances each time.

If you are not able to suspend you have a statutory right to claim interest on the amount unpaid under the Late Payment of Commercial Debts (Interest) Act. This allows for a statutory rate of interest or, subject to it being a substantial remedy, such other rate as the parties agree.

T is for termination

Termination occurs when:

- Everything that needs doing under a contract is done (discharge by performance) or nothing more can be done due to an external force (frustration);
- Both parties agree;
- A breach occurs that triggers the common law right of repudiation (discussed before) or a contractual right to terminate.

Here we will look at contractual termination rights in more detail, as they are common in construction contracts. It is worth noting that the termination can operate on different levels – often it will be the termination of the employment of a party (meaning that the party doing works or performing services is no longer obliged to carry on with them) which actually leaves other obligations (such as those relating to confidentiality) in place. Other times it might be ending the contract itself. Here we are considering 'termination' to mean 'termination of employment'.

Some contracts have what is called termination 'at will' or 'for convenience', where the party that has that right can simply end the contract by giving notice to the other party. This is usually to allow it an exit route due to other circumstances and will happen more commonly before a project starts on site (so a consultant may be terminated when the client is not able to acquire the land needed for the project) but may occur afterwards when the client no longer requires the project to be completed (such as a client mothballing a development indefinitely due to market conditions). It is usual for the party that has the right to pay for the privilege because it allows flexibility.

Many contracts allow termination for actual (or, occasionally, even the prospect of) insolvency of the parties. If the contract does not allow for insolvency, the insolvent party is likely soon to be in breach of contract for non-performance.

Termination for breach can be triggered by reference to:

- Broad standards of performance, where the failing is usually something such as 'material breach' or 'persistent breach';
- Set criteria, such as reaching a defined percentage (usually a figure up to 100 per cent) of any liability cap set out in the contract.

Termination for breach (which is often defined as including insolvency) will have a mechanism for how moneys may be retained by the innocent party and/or paid to the party in breach, should there be any left after the costs of termination have been calculated.

9.5 Summary and tutorial questions

9.5.1 Summary

As stated at the beginning of the chapter, suspension and termination of a construction contract are serious steps and should only be considered in exceptional circumstances and when all other avenues have been explored. Suspension is clearly not the end of the contract and is used as a break which is forced on the parties due to unforeseen circumstances (e.g. a major redesign or the employer requires a major financial restructuring of their business). The project will be able to restart after the suspension has been lifted, but there will necessarily be additional costs due to remobilisation, extensions of time etc. The contract conditions will include a timetable for when the appropriate notices must be given.

If the work is suspended the contractor should expect that:

- They will not receive any payment whatsoever for a period up to 112 days (depending on the contract).
- They will be required to maintain the plant, personnel and subcontractors in readiness to resume the work.
- They will still be responsible for protecting and securing the work already performed and for subsequent repairs, as well as being responsible for securing and insuring the site.
- They may have to bear the financial burden of these obligations depending on the contract conditions – FIDIC allows the contractor to be recompensed under Clause 8.9(b) but this clause is invariably deleted by employers.

Termination is a far more permanent state of affairs and the employer generally has the right to terminate 'for cause', i.e. because the contractor has failed in some major way or external circumstances have affected the project so that it is impossible to continue. The 'causes' are normally clearly set out in the contract. The employer may also have the right to terminate 'for convenience' so that no cause needs to be proved, but the market conditions may have altered sufficiently for the project to have become unviable. In this case, the building will normally be finished to a particular point (e.g. waterproof and weather proof) to allow a restart at some point in the future.

Many standard forms of contract also give the contractor the right to suspend or terminate the contract 'for cause', which usually relates to non-payment, failure of the employer to fulfil their obligations, prolonged suspension or employer insolvency, e.g. FIDIC Clause 16; although it is not unusual in international contracts for this entire clause to be deleted by the employer.

9.5.2 Tutorial questions

1. Under what circumstances would an employer wish to suspend the works?
2. Under what circumstances would a contractor wish to suspend the works?
3. Outline the essential difference between termination 'for cause' and termination 'for convenience'.
4. The employer, a government ministry, has signed a contract for construction works with an international contractor, who has commenced mobilisation procedures. Shortly after the contract signing, it transpires that the Ministry of Finance has not (yet) approved a budget for this project, which is a formal requirement for payment. Discuss the contractor's options.
5. Discuss the concept of 'good faith' in relation to termination for convenience.
6. If the employer terminates the contract, can the contractor object using the dispute resolution procedures?

10 Practical or substantial completion, delays and damages

Although the term has been used for decades, there is no precise legal definition of practical or substantial completion and it can have various meanings, ranging from 'nearly but not quite complete' to 'substantially complete' or 'complete for practical purposes'. So basically, for all intents and purposes, the project is complete and the employer or end-user can move in and use it for the purpose they intended when they commissioned the project in the first place. Of course, there will be small snags and minor issues which the contractor is obliged to finish off before the project has achieved 'final completion'. The acid test at the point of substantial completion is therefore – can the building or facility be used for the intended purpose?

The confusingly different terms arise from the different titles in the various standard forms of contract – 'practical completion' in JCT contracts, 'substantial completion' in ICC contracts, plain old 'completion' in the NEC, and 'taking over' in FIDIC contracts. Whatever title is used, this event will trigger the following:

- Release of normally half of any retained monies (the remainder being released when the list of defects has been satisfactorily completed).
- The insurance obligations and risk of loss or damage to the works passes from the contractor to the employer.
- The contractor is no longer liable for any liquidated damages or penalties.
- The defects correction period begins at this point.
- No further instructions or variations can be given to the contractor.
- The start date for the statute of limitations for any contractual claims.
- Performance bonds and the like are not required and should therefore be terminated.

10.1 Completion under the standard forms of contract

In the JCT forms, except the Major Project Construction Contract, there is no definition of practical completion at all. Instead, discretion is provided to the contract administrator to determine whether the works have achieved this level of completion. The infrastructure standard forms also adopt a minimalist approach, by requiring 'substantial completion', which, again, is not a defined term. The drafters of NEC3 have at least tried to provide more clarity by abandoning the preliminary adjective and opting simply for 'completion'. This requires the contractor to carry out all work within the 'works information' and rectify any defects which would have prevented use. This on its own could possibly mean defect-free completion of all that is in the works information. The Third Edition of the NEC added a final paragraph which suggests that if there is no specific provision in the works information about completion then the test is based on whether the contractor has done all the work necessary for the employer or end-user to use the completed facility for whatever they want to do in it (see paragraph above). It is therefore possible for the works to contain minor defects and still be usable, and so it is therefore complete. Glad that's clear then.

Most of the case law in the UK deals with practical completion under the JCT Conditions of Contract, so it is of limited help when it comes to the ICC or NEC definition and even less of a help in international contracts. From the cases under English law, it is possible to say that the indicators of practical completion are:

- The construction work to be done under the contract must be complete.
- There must be no apparent defects which prevent the facility being used.
- Any minor defects should be able to be put right without undue interference or disturbance to the building occupier. This is also referred to as the *de minimis* principle.
- If the snagging list is excessively long, practical completion should not be given, even though each individual snag may be relatively minor.

However, practical completion is possible if there are hidden or 'latent' defects or, as stated above, if minor work remains to be finished off. The Defects Liability Period (the term used in JCT contracts) enables previously undiscovered defects to be remedied. However, the Certificate of Practical Completion may not be issued if there are patent defects, which means that the works have not yet been satisfactorily constructed.

The above is therefore a useful guide, but not an unambiguous definition of practical completion – as we've seen, there isn't one. This is not surprising as it is an issue which can depend on the facts in each project; so a simple solution would be to define completion at the beginning of the project in the contract agreement. If certain elements are required, then they should be clearly defined as a deliverable in the contractual scope of works in order to avoid any misunderstandings or confrontation later on.

Interestingly, if the employer unilaterally decides to take possession of the building or facility before the Certificate of Practical Completion has been issued, then practical completion will be deemed to have occurred – so employers beware. This issue was the subject of an English legal case in the 1970s.

10.2 Practical / substantial completion and the date for completion

The discussion above purely concerns itself with the physical event of completing the works and all that it entails. However, the contractor is contractually obliged to reach this event by the completion date mentioned in the contract details. If they don't achieve practical / substantial completion by this date and overrun, they risk being charged the liquidated damages or penalties stated in the contract. If they finish earlier than the completion date, this gives rise to 'contract float' with its own issues and difficulties – clearly the contractor will wish to boast about finishing early, but the employer is still within their contractual rights to issue variations or additions, provided they have not yet issued the Certificate of Completion. These problems are only created when the float amounts to a substantial amount of time and most contractors will wish to factor in a margin of safety to ensure that they achieve substantial completion by the due contract completion date.

In the circumstances where the contractor fails to complete the works by the completion date, this will generate a claim for liquidated damages from the employer, as the contractor is late in handing over the works and therefore in breach of contract. At this point, the PM / engineer, through the contract administrator, must issue a Certificate of Non-Completion as well as ensuring that any requests for extensions of time from the contractor have been fully addressed. In the event the employer wishes to deduct liquidated damages from any sums due to the contractor, a notice must be given to the contractor stating an intention to do so, which must be given no later than the date of issue of the Final Certificate, although in most instances that date is some time away.

A Certificate of Practical or Substantial Completion does not (and is not intended to) offer end-users, tenants or purchasers (including subsequent purchasers) or their lenders any direct reassurance that the works have been carried out to any other standard than that laid out in the construction contract. It is purely part of the internal contractual machinery under the building contract which regulates the rights and obligations between the employer and the contractor. It is issued to identify that the works are practically complete and to trigger a further chain of events, as described above.

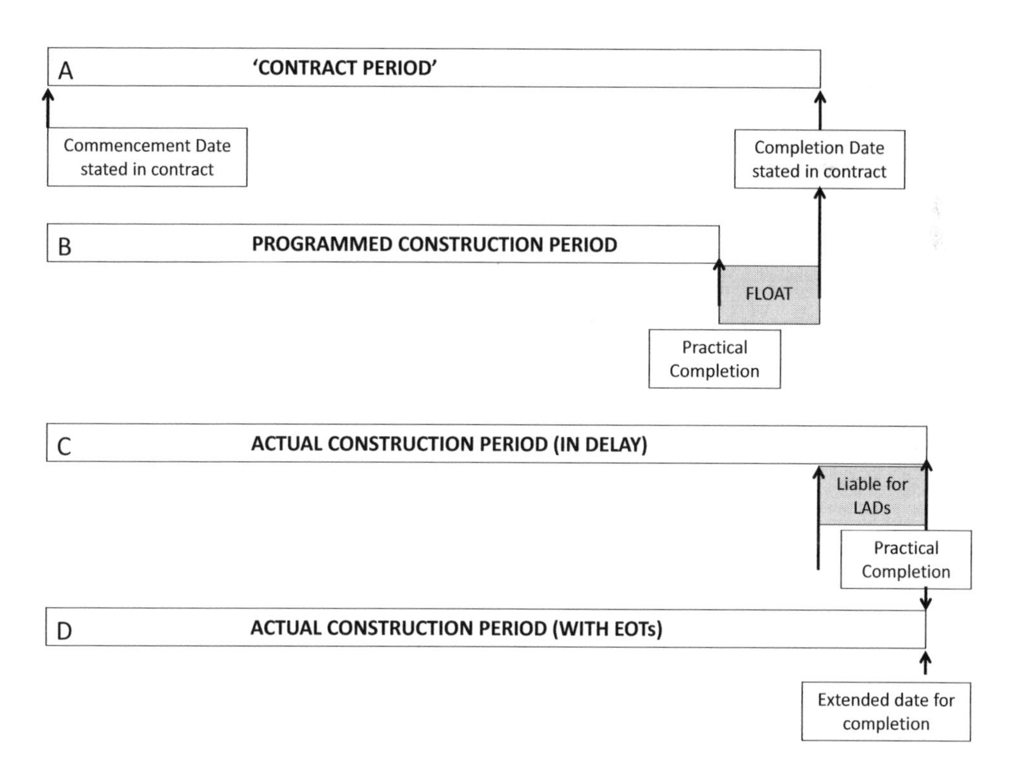

Figure 10.1 Substantial completion vs. contract date for completion.

10.3 Delays and extensions of time

Delay is defined as the time overrun, either beyond the date for completion as specified in the contract documents, or beyond the date that the parties have agreed upon following any extensions of time issued on behalf of the employer. Scenario C of figure 10.1 on page 145 illustrates the delay situation, although this diagram only illustrates that the total project is in delay and as the project is made up of individual operations all linked together in a critical path network, it is clear to see how this gets much more complicated.

Delay is effectively a project slipping over its planned schedule and is a common problem in construction projects for a variety of reasons. To the employer or other end-user of the building, any delay means a potential loss of revenue through lack of production facilities or rentable space. To the contractor, a delay may mean increased overhead costs due to the extended work period, possible material costs through inflation, and increased labour costs due to reduced productivity together with not being able to start the next project as early as they would wish. Completing projects on time is an indicator of efficiency, both for the project manager and the contractor, but the construction process is subject to many variables and unpredictable factors, which are generated from many sources, including the performance of parties, resources availability, environmental conditions, involvement of other parties, and contractual relations. All of these factors conspire to affect the efficient and effective running of the project, therefore to ensure that the project finishes on time requires constant and proactive management as discussed in detail in Chapter 4.

10.3.1 Main causes of delay in construction projects

The most frequent causes of delay often depend on who you ask. From the employer's point of view, delays are related to both contractor and labour issues. Many employers do realise that awarding contracts to the lowest bidder is an important factor in project delay as the contractor tries to juggle their resources and productivity within the financial constraints that have enabled them to win the job in the first place. Contractors, on the other hand, indicate that the most frequent causes of delay are (not surprisingly) related to the owners – with constant interference, requests for additional works and late payments. Consultants, like owners, consider that the contractor is mainly responsible for project delays as they are the party who generally do the programming and scheduling.

The most common causes of delay as seen by the owners are as follows:

- shortage of competent labour
- unqualified work force
- inadequate experience of contractor
- difficulties in project financing by the contractor
- ineffective project planning and scheduling by the contractor
- low productivity level of labour force
- rework due to errors during construction
- delay in progress payments by employer
- original contract duration is too short.

Similar to employers, consultants often indicate that the most common causes of delay are related to issues under the control of contractors:

- difficulties in project financing by the contractor
- inadequate experience of the contractor
- shortage of competent labour
- delay in progress payments by employer
- delay in material delivery
- poor site management and supervision by the contractor
- ineffective planning and scheduling of project by the contractor
- incorrect procurement
- poor quality of contractor's technical and supervisory staff.

The top ten factors as far as the contractor is aware are:

- delay in honouring Payment Certificates
- poor supervision (from the consultant)
- difficulty in accessing bank credit
- shortage of materials
- underestimation of cost of project
- late delivery of materials
- delay in instructions from consultants
- underestimation of complexity of project
- price inflation
- construction methods.

The above lists have been collated from various studies into the causes of delay in construction projects.

10.3.2 Identifying and analysing delays

Almost all modern construction projects use some form of computerised CPM (critical path method) programming, which allows for systematic planning and control of construction resources. CPM is the planning technique most commonly used in the construction industry today, and is based on the same critical path analysis principles established when it was invented, but with the added benefit of using modern computer software, such as Primavera or Asta Powerproject.

A significant aspect of delay analysis is the interrogation of records upon which the original baseline schedule was developed as well as the as-built records which show the delay in question. Therefore, the need for good record keeping and document control is paramount to a successful delay evaluation.

The identification and analysis of delay entitlement can be both difficult and time consuming. Complex construction projects can have literally thousands of operations being carried out simultaneously and often the delay analysis is carried out by staff who are also trying to manage the progress of the project and may be unfamiliar with forensic analysis techniques or programming skills. This is one of the main reasons why the contractor's applications for delay and extensions of time are often rejected by the employer's representatives – because they have not prepared the application thoroughly enough.

The construction phase is also the phase in which poor design decisions, or lack of sufficient preconstruction planning, will often create critical delays to project completion, as measured by delays to site operations.

10.3.3 Delay entitlement and extensions of time, loss and expense

A valid extension of time application should adequately establish both causation and liability, i.e. what caused the delay and who was responsible. The purpose of delay analysis is to satisfy the causation requirement in such a way that it can be used to assess the resulting loss and expense to the contractor.

As discussed in earlier chapters, the contractor must manage two issues. Firstly, there is the physical progress of the works, which is where the CPM techniques come in and any delay will be analysed by comparing what should have happened with what did happen (Baseline schedule vs. Actual). Secondly, the contractor must also manage the contract and in that contract there is usually a list of 'relevant events' for which the contractor will be entitled to an extension of time, also termed excusable delays – see below. There will also be a separate list of events for which the contractor is entitled to be paid their loss and expense – or as the NEC puts it, compensation events (hence compensable delays as below). The PM / engineer must also manage both of these issues, which will be delegated to the supervision consultants for physical progress and the contract administrator for contract administration.

Delays may be therefore categorised as:

- *excusable*: i.e. delays that are unforeseeable and beyond the control of the contractor;
- *non-excusable*: i.e. delays that are foreseeable or within the contractor's control;
- *compensable*: i.e. delays for which the contractor will receive payment for loss and expense;
- *non-compensable*: i.e. delays for which the contractor will not be entitled to receive payment for loss and expense.

When demonstrating that a delay is both excusable and compensable, the delay must be shown to be critical, i.e. that any delay to this activity will have an effect on the total project duration.

Although it is not in the remit of this book to explain critical path assessment techniques, it is worth noting how a critical activity can be recognised.

When the project has been fully programmed / scheduled by the project planners, there will be a critical path network of all the activities and operations. All construction activities will have a duration and a construction logic, i.e. as to which activity must happen before something else etc. When the process has been completed and the whole project fitted in to the contractual time available, the start dates and finish dates of each construction activity will be established. Some of these activities will benefit from 'float', i.e. the activity duration is shorter than the time available for it (see figure 10.2), others will be more 'critical' in that there is no such leeway. In these cases, the construction activity *must* start at the earliest start date and *must* finish at the earliest finish date.

Within this critical path network there will be one or more critical paths, where all the activities from project commencement to project completion are critical, have no float and therefore any delay to these activities is bound to delay the entire project.

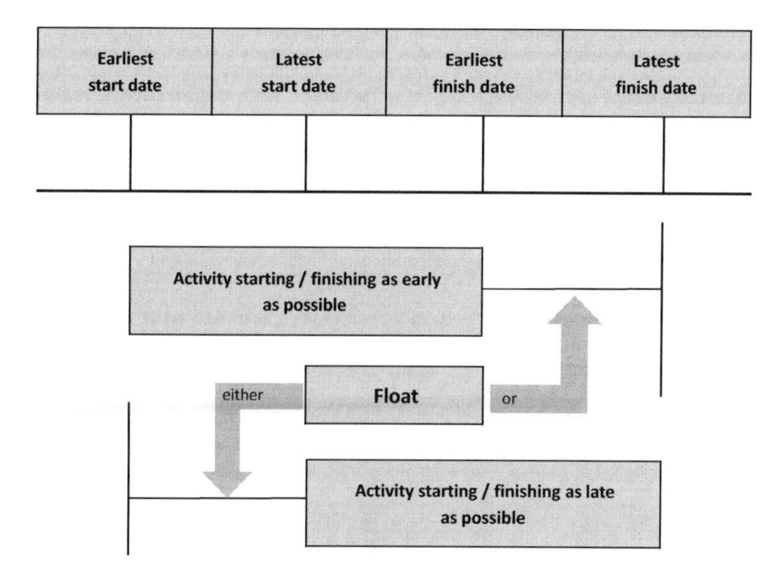

Figure 10.2 Definition of float in a construction activity.

Therefore, a critical activity can be recognised as having no float attached to it and the earliest start date is the same as the latest start date, earliest finish date is the same as the latest finish date.

See also Chapter 13 for a more detailed discussion and analysis of delay and disruption.

The contractor's original programme or baseline schedule is useful evidence of a contractor's original intentions and should serve as the starting point when identifying delays. This is why many engineering contracts insist on the contractor's programme being submitted at the beginning of the project. One of the main objectives of delay analysis is the establishment of a factual timeline of events which actually delayed the project's completion date, which can also help in the preparation and / or validation of the as-built programme. Ideally, an as-built programme will have been prepared and maintained during the progress of the works and, if so, the data used to maintain and update the project programme can also be used when forensically constructing an as-built programme. The process of identifying delay events is a fundamental aspect of delay analysis and can be undertaken in two main ways, either an 'effect-based' approach – i.e. looking at what occurred after the event, or a 'cause-based' approach – i.e. looking at what occurred before the event.

In either way, unless the project has been, or will be, delayed, there is no need for any analysis at all. Extensions of time are for extending the final completion date of the project, so if an activity is delayed which can be taken up by the float, that is normally considered at contractor's risk. Care must be taken, however, that the remainder of that path does not become critical by using up the float.

The other issue which is generally used in the same breath as delay is that of disruption. Delay and disruption often result from the same events, although disruption, unlike delay, always has a financial consequence for the contractor as it can seriously affect their efficient use of resources and the productivity of those resources.

The Society of Construction Law (SCL) Delay and Disruption Protocol defines disruption as:

> loss of productivity, disturbance, hindrance or interruption to a Contractor's normal working methods, resulting in lower efficiency. In the construction context, disrupted work is work that is carried out less efficiently than it would have been, had it not been for the cause of the disruption. If caused by the Employer, it may give rise to a right to compensation either under the contract or as a breach of contract.

Although claims for disruption are unlikely to result in an extension of time, the analysis of loss productivity and efficiency is extremely complex and specialised and should only be considered if the disruption has had a major effect on the project.

10.4 Damages and penalties for non-completion

Despite the best endeavours of all parties, many construction projects still overrun their original contract period for a multitude of reasons. For the employer, a delay will mean that the facility will not be available for use when it was originally intended, which means they may have to incur additional costs or suffer a delay in receiving any rental or other income from the project. For the contractor, delay to the completion of the project means that their resources are tied up for longer than expected and can also result in a liability for penalties or delay damages to be paid to the employer.

As a result of these risks, most building contracts calculate and fix the damages that will be payable by the contractor in the event of late completion, if an extension of time has not already been granted. These are referred to as 'liquidated damages' (LDs) and sometimes as 'liquidated and ascertained damages' (LADs). The term *liquidated* means they are set as a particular sum of money and *ascertained* means they have been pre-calculated as an average assessment of the actual costs of delay to the employer or end-user. The term *penalty* is often used in international contracts and care must be taken when using this term. In most legal jurisdictions, failure to complete a scope of works by the contract completion date is a straightforward breach of contract and the redress for such a breach is damages to be paid by the offending party. Any attempt to *penalise* the offending party would be unlikely to be successful in court if the penalty was to be challenged. Normally, only the official courts of law are allowed to penalise. However, in some cases (especially in oil and gas contracts) the damages or loss of profit which results from a delay to the facility are of such magnitude that imposing the actual loss as damages would more than likely bankrupt the contractor. Therefore, a penalty clause, which may be set at less than the actual loss of profit, is a common feature in such contracts.

10.4.1 The importance of time in construction contracts

Duration is one of the critical features of any contract (along with cost and scope of works – our old friend time–cost–quality). Therefore substantial provisions are made in construction contracts for establishing the date by which a contractor must complete the scope of work that it has agreed to perform. If no date is specified a 'reasonable time for completion' will normally be implied, based on the circumstances of the particular project, which is called 'time at large'. Contracts which have time at large are a nightmare scenario for the employer and should be avoided at all costs. Contractors, on the other hand, are quite happy with time-at-large contracts as they can proceed with the works at a steady pace, although there would be an implied requirement to work diligently and to use their resources efficiently. See also section 13.3.1 for a discussion of time at large.

Without a fixed contract completion date, the employer is clearly unable to charge liquidated damages if the project is late (what is 'late' if there is no completion date?); so for this reason, construction contracts will include a fixed completion date with further provisions for it to be extended under certain circumstances. Extensions of time provisions therefore have benefits for both parties:

From a contractor's perspective:

- They allow additional time to cover delays that are neither the fault nor the responsibility of the contractor, such as *force majeure* or where the employer has created the delay.
- A contractor will not normally be liable to pay damages for delay if completion time has been validly extended.

From an employer's perspective, the ability to extend the completion date of a project:

- maintains an actual date for completion, even when circumstances occur which means the original date cannot be met;
- protects the employer's right to claim liquidated damages for delays where the contractor is culpable.

Therefore time and schedule are critical components in the management and administration of construction contracts. The contract administrator needs to be very clear about the difference between the 'contract period' and the 'construction period' as they are very different concepts. The contract period is the time between the date for commencement and the date for completion as stated in the contract documents. This is the period of time during which the contractor must complete the works. The construction period is the actual programmed duration of the construction schedule and could be shorter or longer than the contract period. If it is shorter, there will be a period of 'float' until the date for completion is reached, and there is considerable argument in the industry regarding this time duration. If the construction period is longer than the contract period, then the contractor is in delay and will be subject to LADs, unless the date for completion has been extended. All standard forms of contract give a list of relevant circumstances which will allow the extension of the completion date. See figure 10.1 at the beginning of this chapter for a graphical representation of these points.

10.4.2 Requirements for a valid liquidated damages provision

The obvious benefit of a valid liquidated damages clause from the employer's perspective is that it removes the need for the employer to prove its actual loss and to 'mitigate', that is, to take reasonable steps to avoid or reduce the loss covered by the

LD clause. In fact, if the clause is properly constructed, the employer will be entitled to the specified amounts even if they have sustained no actual loss. From a contractor's perspective, LDs effectively act as a limit on their liability for delay.

Although courts of law will uphold LD clauses wherever possible, a risk still exists that a poorly drafted clause may be unenforceable. Therefore, it is important to ensure that:

- the clause is a genuine pre-estimate of any loss likely to be sustained, rather than a penalty designed to deter a party from breaching the contract in the first place. As stated above, if the clause amounts to a penalty it is likely that it will be unenforceable;
- there is an actual formula stated in the contract from which damages will be payable;
- any specific contractual procedures, such as giving written notice within specified time limits, have been followed;
- the employer has not waived the right to deduct such damages.

In addition, a party cannot impose a contractual obligation on the other party where it has impeded the other party in the performance of the obligation. This is particularly important where employers are accused of interfering in the progress of the contractor's work and this can be done in many ways. As mentioned in Chapter 1, the employer must allow the contractor to proceed with the works and this 'employer interference' has been successfully used as a defence against the imposition of LDs. Employers beware.

Therefore, to ensure that charging LDs against the contractor is valid, the employer must establish:

- that there has been a breach of contract (i.e. the contractor is in delay);
- that it has suffered loss;
- that there is a causal connection between the breach and the losses that party seeks to recover.

In complex construction contract disputes there can be many losses and breaches of contract which can make it extremely difficult to link a specified loss to a specified breach. Therefore, contractors often submit what are termed 'global claims' where the delay has been caused by a multitude of concurrent factors. This is a nightmare to sort out but provides good business for claims consultants and delay analysts.

10.4.3 Sub-contracts and liquidated damages

To minimise exposure, many contractors will seek to enter into contracts with the sub-contractors on substantially the same terms as their contract with the employer as discussed in Chapter 7. Stepping down the main contract terms works well for most obligations, however attempting to pass down liability for LDs to sub-contractors can be difficult. The contractor's loss will have three main elements:

- its own costs of delay - site costs, overheads, financing etc;
- its liability for LDs under the main contract;
- its other sub-contractors' loss and expense.

The first two are fairly easy to estimate, but the effect of one sub-contractor's delay on the progress and cost of the others is almost impossible to pre-estimate. A number of the sub-contract standard forms therefore avoid LD clauses, instead providing that the sub-contractor must pay the contractor for its loss and expense resulting from the sub-contractor's failure to complete on time. This loss and expense may include all three elements outlined above.

10.5 Early use, partial possession and sectional completion

Early use occurs when the employer, *with the consent of the contractor*, uses or occupies part of the works before that part is deemed to be completed. Early use will leave the risk and responsibility for the site and the works with the contractor, and therefore the contractor's full liability for penalties and damages as stated in the contract will continue. However, if the employer delays or disrupts the contractor, this may give rise to a claim for loss and expense and possibly extension of time. This is clearly not a very satisfactory arrangement from a contractual point of view and should only be considered as an emergency arrangement.

Partial possession of a project is where an employer is allowed, either by pre-arrangement in the original contract or with the express consent of the contractor, to take possession of part of the works, while the contractor is still completing the remainder. If the employer exercises this right then that part of the work is effectively deemed to be completed and the contractor will therefore hand over that portion which is now the responsibility of the employer. Most standard forms of construction contract have clauses to reduce a contractor's liability for penalties or liquidated damages proportionately if any part of the works is certified as complete before the contract is completed in its entirety.

Sectional completion is best used where the employer requires the contractor to complete a defined part of the works so that the employer (or possibly another contractor) can take over that section entirely and either use the section or carry out further works (for example, the bedrooms wing of a hotel project may be handed over before the full hotel is complete in order for the rooms to be furnished by others). All commercial projects, particularly large ones, may reach practical completion in stages – for example, as stated, the bedrooms of a hotel may be completed and ready for furnishing before the main kitchens have been completed; this would allow the project to be completed in sections, without the employer or end-user taking full possession. The contractor would not then be expected to carry out any further work to that section, although it may remain liable for defects, depending on the status of the appropriate Sectional Completion Certificate.

In cases of sectional completion, care must be taken by the contractor, as practical / substantial completion of the appropriate section occurs when that section is complete, but the defects correction period of the section will run until the end of the full project, as shown in figure 10.3. In this case the defects correction period of section 1 (S1) is substantially longer than the period for S2, which in turn is longer than the 'normal' defects correction period of the whole project. This means that, although the contractor will be repaid a proportion of the retention fund and the employer cannot issue Variation Orders for this section, the contractor will still be responsible for snagging any defects which manifest themselves – while other contractors may be working in the area.

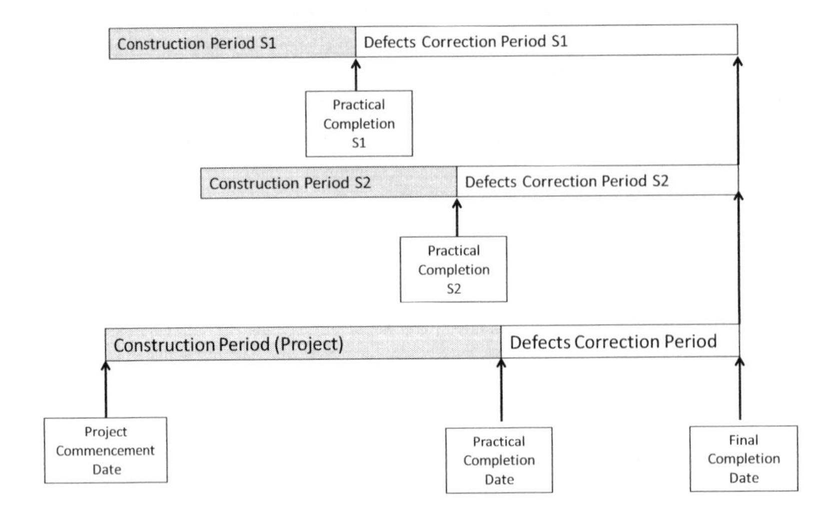

Figure 10.3 Defects correction period and sectional completions.

The benefit to the contractor of sectional completion is that it is a microcosm of practical completion in that it is relieved of some of its contractual obligations, including insurance of the works and a relevant proportion of liquidated damages. As stated above, the contractor may also be entitled to repayment of a proportional part of the retention monies.

Most standard forms of contract contain a right to have the works completed in sections, and a corresponding right for the contractor to expect a reduction in its liability for liquidated damages.

Calculating LDs for sectional completion should be approached with even more caution than single LDs:

- The sections should be clearly understood by all parties, and if possible the terminology used to refer to them should be consistent throughout the document.
- Separate dates for completion should be given for each section.
- Separate LDs should be stated for each section so that, when sectional completion is reached, the contractor's LD liability may be proportionately reduced accordingly.
- The contractor's liabilities for failure to meet sectional completion dates should be made clear.

Where an employer wishes a project to be completed and handed over in sections, and this is addressed in the contract with a provision for different liquidated damages, care should also be taken when the commencement date of one section depends on the completion of a previous section. A delay by the contractor in an early section may not automatically generate grounds for an extension of time for subsequent sections; therefore, in the case above, a delay in handing over the bedrooms wing of a hotel may not be sufficient grounds for an extension of time for the furnishings contractor. But, as ever, the terms and conditions of the various contracts between the parties would decide the issue.

10.6 Summary and tutorial questions

10.6.1 Summary

Practical / substantial completion is an important event / milestone in a construction contract, which gives rise to a number of further rights and responsibilities for both the employer and the contractor. It is therefore important to ensure that all parties have a clear understanding of the criteria which must be satisfied at this event together with the rights and responsibilities this represents.

Construction projects are complex operations and the building can be usable before all construction is finished. There will always be minor snags or small items of work to finish off and as long as these snags do not affect the use of the building, there is no reason why the building cannot be handed over whilst the snags are being rectified. The Certificate of Substantial Completion will be issued by the PM / engineer together with the snag list / punch list and this triggers the events given at the beginning of the chapter. This event should occur by the contract completion date as stated in the contract – if it doesn't, the contractor is in delay and will be liable for LDs, unless the completion date has been extended by an EOT.

Delays to the contractor's progress are caused by a multitude of reasons; some are caused by the employer, some by the contractor and some by external factors beyond the control of either party. Many standard forms of contract contain a list of 'relevant events' whereby the contractor will be awarded additional time or additional costs in compensation for the delay. As a general rule of thumb, an extension of time is awarded for employer culpable delay or *force majeure* and additional costs are payable only for employer culpable delay – see table 10.1. As this is a general rule, the detailed requirements will differ from project to project.

However, just because there is a delay to an activity doesn't mean that the project will be delayed, as there may be sufficient float in the programme to absorb the delay and an extension of time is an extension to the overall contract completion date – therefore the delay event must be analysed to assess its effect on the project as a whole. The various techniques of analysis are discussed more fully in Chapter 13.

As stated above, construction projects are complex and may be made up of different sections, parts, phases etc. Therefore, it may be useful to complete and hand over each section separately to allow the employer to progressively take over the works, rather than all in one go. To enable this to happen, many standard forms of contract allow early use, partial possession or sectional completion which effectively creates several mini projects for the completion and close-out procedures: although contractors should be aware that the defects correction period for the early sections may be considerably longer than the later sections. This is due to there being one final completion date for the entire project.

Table 10.1 General rule of entitlement for extension of time and additional costs

	Contractor culpable delay	*Employer culpable delay*	*Force majeure*
Extension of time	X	✓	✓
Additional costs (loss and expense)	X	✓	X

10.6.2 Tutorial questions

1 Construct an effective working definition of substantial completion in a construction project.
2 What events are triggered by the issue of the Certificate of Substantial Completion?
3 Why is it important for both the employer and contractor to be able to extend the contract completion date?
4 If the Certificate of Substantial Completion is issued after the contract date for completion, should the employer automatically charge liquidated damages to the contractor?
5 If an employer culpable delay occurred at the same time as a contractor culpable delay, should an extension of time be granted? (This question may be reviewed after reading chapter 13.)
6 If a contractor 'designs in' float on a construction programme, should this float be maintained when considering extensions of time?

11 Defects correction, final completion and close-out

11.1 Defects liability and correction

Chapter 10 discussed the issue of substantial completion, i.e. the point when the contractor's work is finished, apart from rectifying minor snags and defects in materials and workmanship found during the inspections by the supervision consultants. Sometimes the contractor is also required to repair or maintain the works for a certain period after completion and this period of time is variously referred to as the *Defects Liability Period* (in JCT contracts pre-2005), *Rectification Period* (JCT contracts post-2005), *Defects Correction Period* (ICC and NEC contracts) and *Defects Notification Period* (FIDIC contracts). Some standard forms of contract also use the terms *Guarantee Period* or *Warranty Period*, which can be easily confused with the guarantee / warranty period given by equipment manufacturers and therefore should be avoided. It is slightly disappointing that all the main standard forms of contract use different terminology for what is in principle an identical concept.

For simplicity and clarity, this book will use the term *Defects Correction Period*, as this term best defines the formal requirements, which is the period of time when the contractor must repair and correct any defects in the works identified by the supervision consultants as stated above. All expenses in correcting the defects will be borne by the contractor and no additional costs should be charged to the employer (as the contractor is deemed responsible for the defective work). It is a period usually between 6 and 12 months following practical / substantial completion, during which time the employer can also require the contractor to return to the site to complete any omissions in the works and, as stated above, to make good any defective work or materials. When all omissions and defects have been made good, a Certificate or Statement of Completion of Making Good Defects will be issued, which is a pre-requisite to the final Certificate of Completion.

The effect of the defects liability and correction provisions is not only to confirm that the contractor is liable and obliged to correct the defects (this liability will exist in tort under most legal jurisdictions anyway) but also to give the contractor the right to receive a notice of any defects from the employer's representatives within the stipulated period and to have the opportunity of correcting the defects at their own expense. The commencement of the defects correction period is the date of practical / substantial completion, i.e. the point when the employer takes over the works and 'moves in', unless otherwise agreed by both parties.

11.1.1 What is a defect?

A defect in this context is a failure of some aspect of the completed project to satisfy the express or implied quality or quantity obligations of the construction contract. Typically, there are four categories of defects in a construction project:

- design deficiencies
- material deficiencies
- construction deficiencies, and
- sub-surface / geotechnical problems.

Clearly, the contractor cannot be held responsible for design deficiencies, unless the project was based on a design–build arrangement. (For example – insufficient reinforcement in a concrete beam may cause cracks to appear as the beam buckles under the weight of the load from above. The structural engineer who designed and specified the reinforcement in the beam should take responsibility for this issue.) Therefore, to require the contractor to remedy and pay for defects caused by design problems would be unfair, although the contractor would no doubt try to help in any way they could. This would be a very difficult issue to resolve legally as it would involve claims against the professional indemnity insurance of the design engineer and there is also the point that the contractor should have sufficient expertise to know whether a structural design is defective. A major word of caution at this point concerns the modern trend for employers to insert a 'design verification' clause into the construction contract. These clauses have the effect of requiring the contractor to verify the suitability, and possibly completeness, of the design issued to them. A very dangerous 'stealth' clause, of which contractors are well advised to be very wary.

Similarly, material deficiencies would be covered under the manufacturer's warranty but once they have been incorporated into the works, it may be extremely difficult to take the materials out and replace (for example, facing bricks where the facing material falls off – termed spalling – would be virtually impossible to replace once the wall is built especially if it is a load bearing wall). Also, poor quality aggregate in concrete would be physically impossible to take out after the concrete is placed. Additionally, the contractor is expected to reject unsuitable materials and on many international projects there is a 'Materials Approval System' whereby the supervision consultants are required to approve the procurement and delivery of all major materials, thus partially exempting the contractor from that responsibility.

Conversely, defects appearing before substantial completion are also required to be rectified by the contractor but the main difference is that before substantial completion, the PM / engineer can issue instructions and variations to the contractor – after this point, the PM / engineer's instructions can only be to rectify defects – not to add or omit work as a variation to the scope of works.

11.1.2 Patent and latent defects

Patent defects are 'obvious' defects, whereas latent defects are hidden and only become apparent at some later date. A patent defect is therefore a problem which is visually obvious at the time of inspection, for example, the omission of mastic seal around a shower or bath unit, and will be recorded in the snagging or defect lists during the inspection prior to substantial completion. If defects become patent during the defects correction period, most standard form of construction contracts contain mechanisms for rectifying them. Latent defects may include those issues that, while not obvious at the time of inspection, will become obvious after a period of time or usage of the building or even after an event such as heavy rain which can expose poor construction damp proofing or insufficient roof overflow. There may also be defects that are not discovered because the person charged with discovering them failed to carry out the investigations properly; so for this reason, English law defines latent defects as those defects that do not become obvious even though the requisite level of skill and care had been exercised in searching for and identifying them.

11.1.3 Liability for defects

It is worth mentioning that the liability of a contractor for defects is generally wider than purely to the employer under the construction contract. This means that third parties who have acquired an interest in the facility which has been built may be able to claim damages for a breach of contract caused by the defect if they have suffered loss and may also be able to sue the contractor in tort. As mentioned in the Preface, this is not a textbook on construction law, but contractors need to be aware that their liability for defects is not only contractual and not only to the employer. See discussion on rights of third parties in Chapter 7, section 7.3.5.

During the defects correction period, all standard forms of contract give the contractor permission to return to the site for the purpose of remedying the defects, which is in effect a right to repair or make good defective works. The standard forms of contract also envisage that defects might manifest themselves during the defects correction period itself (i.e. after the initial snagging list has been produced and transmitted to the contractor) and such defects will not be considered as a breach of contract. Upon the receipt of all notices to rectify defects the contractor is obliged to return to the site to make good the defects and, if they fail to do so, they are clearly in breach of contract and the employer may entitled to be paid damages to recover the cost of remedying the defect by others.

Furthermore, the contractual liability of the contractor is not limited to the duration of the defects correction period and even though a Certificate of Making Good Defects may have been issued by the PM / engineer, the contractor's liability for not completing the works in accordance with the contract continues until the time limits given in the statute of limitations of the governing law of the contract. In the UK, this will extend the liability for a period of 6 years for a simple contract, and 12 years for a deed, from the date when the cause of action arose. Other legal jurisdictions may have different time limits.

The contractor's liability can also extend beyond the defects correction period where there is an express or implied warranty as to fitness for purpose. In general, an employer under a construction contract does not warrant the feasibility of any design set out in the contract documents. Similarly, an architect or engineer does not warrant to the contractor or builder the buildability of the plans, drawings and specifications prepared by them, or of any temporary works, if indicated in the specification. The contractor is therefore well advised to investigate these matters for themselves and should not rely on the drawings and specifications. Traditionally, when an employer engages a contractor to construct a building on the basis that the building will be constructed in accordance with an architect's design supplied by the employer, the contractor, whilst accepting to carry out the works in accordance with the design documents, makes no promise that the building will fulfil its intended purpose, except where this is specifically written into the contract.

Nonetheless, some limited design responsibility may be placed on a contractor outside that of the total package deal such as design–build contracts. For example, oil and gas contracts invariably use procurement arrangements whereby the contractor completes the working drawings from an outline or scheme design (termed FEED – front end engineering and design). In this case, the contractor has responsibility for choice of materials and, consequently, for defects arising from an incorrect choice – even though they may have the employer's approval for the materials through a materials approval system (MAS). Therefore, by virtue of design documents failing to specify all materials, a choice of materials is left to the skill and judgement of the contractor and this is a rich vein for disputes affecting liability for defects.

It is therefore in the interest of a contractor to carefully consider the implications of the design, even where they have no strict design responsibility. In some jurisdictions as well as some forms of contract, a contractor is under a duty to warn the employer of any problems whatsoever, including design issues, which they should be able to foresee given their expertise. Clearly, in a design–build contract, the contractor will be under an obligation to ensure that the finished product will be reasonably 'fit for its intended purpose' but even when the contract is not a design–build but it is reasonable to imply into the contract a fitness for purpose term, a prudent contractor would be well advised to alert the employer to any obvious design defects that they come across. This is especially important in common law jurisdictions, such as the UK and USA, where the fitness for purpose duty is often stricter than the ordinary responsibility of a designer to exercise due care, skill and diligence.

11.1.4 Snagging lists and punch lists

'Snagging' is the term which is used to refer to a process that takes place a week or so before substantial completion when the works (or a section of the works if the contract allows for sectional completion) is considered complete by the contractor and offered for inspection.

A 'snagging' list is then prepared by the PM / engineer, as advised by any separate supervision consultants, often in conjunction with the contractor. Any major snags must be corrected before the issue of the Certificate of Substantial Completion, but minor *de minimis* snags can be rectified by the contractor during the defects correction period. The actual term 'snagging list' has no universally agreed meaning in the construction industry and is not referred at all in the JCT suite of contracts, but custom and practice have taken it to mean:

- a pre-completion list of outstanding work, which is shared with the building contractor to clarify what work remains outstanding before final completion can be certified;
- a list of minor items of work that remain to be completed at the date of substantial completion.

Although not expressly provided for in most forms of contract, the courts have long recognised the necessity to be able to certify substantial completion subject to minor items which will not affect the employer's ability to take over the works. If the contract does not expressly refer to such a list, it is good practice to make appropriate allowance for a snagging list in a contract amendment.

As mentioned above, major items of work still to be carried out by the contractor should not be included in a snagging list, as substantial completion should not be certified before these items have been completed. Once the Certificate of Substantial Completion has been issued, all major items of work are deemed to be completed.

A 'punch list' is slightly different, in that it will be prepared by the contractor and lists defects or minor additional works which they know require completion or correction. This list will be forwarded to the PM / engineer for their review, amendment and issue. Clearly, this is only a minor procedural difference and for all practical purposes, a snagging list and punch list are the same.

The snagging inspection is a formal event in the management of the contract and should take place following all necessary testing and commissioning of installations and services etc., the removal of all protective materials, final clean-up and full operation of permanent lighting, power, MEP, HVAC etc. On large projects the inspection process may need to be carried out in sections as areas are progressively completed, cleaned up and closed off, which means that snagging on these types of projects may begin months before the overall project completion. Any such areas are then subject to a rigorous locked-access regime to prevent deterioration and damage and will also require a further final inspection just prior to final handover.

It is vital to ensure that meticulous dated and signed records are kept and all receipts filed. Photographic evidence can be a particularly useful record as defects invariably become the subject of disputes long after completion and occupants can also cause their own damage after handover, which they may try to blame on the building contractor.

Snagging / punch lists are therefore a fact of life in all construction projects and make practical and commercial sense to both the contractor and employer. At best, they provide a sound means of communication between the employer, contractor and PM / engineer at the end of the construction period, facilitating an orderly transition between the construction stage and its final use. Once the list of snags has been generated, the contractor's satisfactory completion of the snags clearly needs to be recorded and signed off, so that the Certificate of Making Good can be issued. This is carried out by means of a 'Snagging Log' as shown in table 11.1.

11.1.5 Testing and commissioning

Many parts of a building project require testing during the construction process. For example, the drain runs between manholes will be tested immediately after construction to ensure that they are set at the correct gradient and do not leak. This is clearly only part of the drainage installation and at the end of the project, when everything has been built and installed, the whole system will be tested to make sure it does what it is supposed to do and satisfies all necessary quality standards including the project specification and any relevant statutory regulations.

Commissioning refers to the process of bringing an item into operation and ensuring that it works. This therefore relates more to MEP services which are taken from a

Table 11.1 Example of a Project Snagging Log

PROJECT SNAGGING LOG								
Area	Location	Snag Ref. No.	Date Snag generated	PM / Eng. responsible	First Inspection		2nd Inspection	
					Date checked	PM / Eng. remarks	Date checked	PM / Eng. remarks

state of static completion (putting all the bits together) to a system or installation in full working order. For each system, the process of commissioning will include switching on, regulating and calibrating to achieve optimum performance, testing any control systems and recording the settings for future operation and maintenance staff – normally by creating and handing over an operation and maintenance (O&M) manual.

Building services in a construction project which will require commissioning before the Certificate of Substantial Completion can be issued may include:

- HVAC systems (heating, ventilating and air conditioning)
- generators
- switchboards
- water supply and sanitation
- pumps
- motors
- fire detection and protection systems
- information and communications technology (ICT) systems
- security systems
- facilities management systems
- process plants
- lifts, elevators and escalators.

If the project is for the construction of a process engineering facility or a turnkey contract which includes the equipment within the building as well as the building itself, the commissioning stage is vitally important to establish practical or substantial completion. Therefore, the contract documents must clearly state:

- Who will be responsible for commissioning the various elements?
- What methods, standards and codes of practice are to be used in the process of commissioning?
- Who will accept or 'sign off' the results and what will happen to them?

Figure 11.1 shows a standard flowchart for the procedure required in project testing and commissioning.

11.2 Final completion

When all the defects required to be made good by the contractor have been completed, the PM / engineer will issue a Certificate of Making Good Defects (now just called a Certificate of Making Good). This means that the contractor has fully completed all their contractual obligations and can go home with the remainder of the retention money, which must be released at this point. Well, not quite so fast. Although the retention money is due be released at this point, the PM / engineer or cost consultant must also calculate the value of the final account to be included in the Final Certificate. This final account will include:

a all adjustments to the contract sum including variations, instructions and provisional sums;
b agreed figures for all claims and applications for additional payments.

Where formal proceedings have been commenced in relation to any disputes, or even if any disputes are still open, the Final Certificate cannot be issued until these disputes have been concluded. In addition, the Final Certificate itself may be disputed within a certain time limit which should be stated in the contract data. As discussed previously with Interim Certificates, if the employer intends to pay less than the amount certified, they must first issue a Withholding / Pay Less Notice as required under the Construction Act (under English law).

In construction management contracts or management contracting, where there are a number of separate work package contracts or trade contracts, final statements will be issued for each trade contract, with the construction manager / management contractor coordinating the preparation of the Final Report and Contract Close-out Report for the whole project.

11.3 Contract close-out

The term contract close-out refers to the completion and settlement of all the contract terms and conditions, including full completion of the project scope of works by the contractor and full and final settlement of all payments by the employer. The building will now have been fully integrated into the employer's business or whatever else they intended to do with it. All issues such as disputes, variations, changes etc. will have

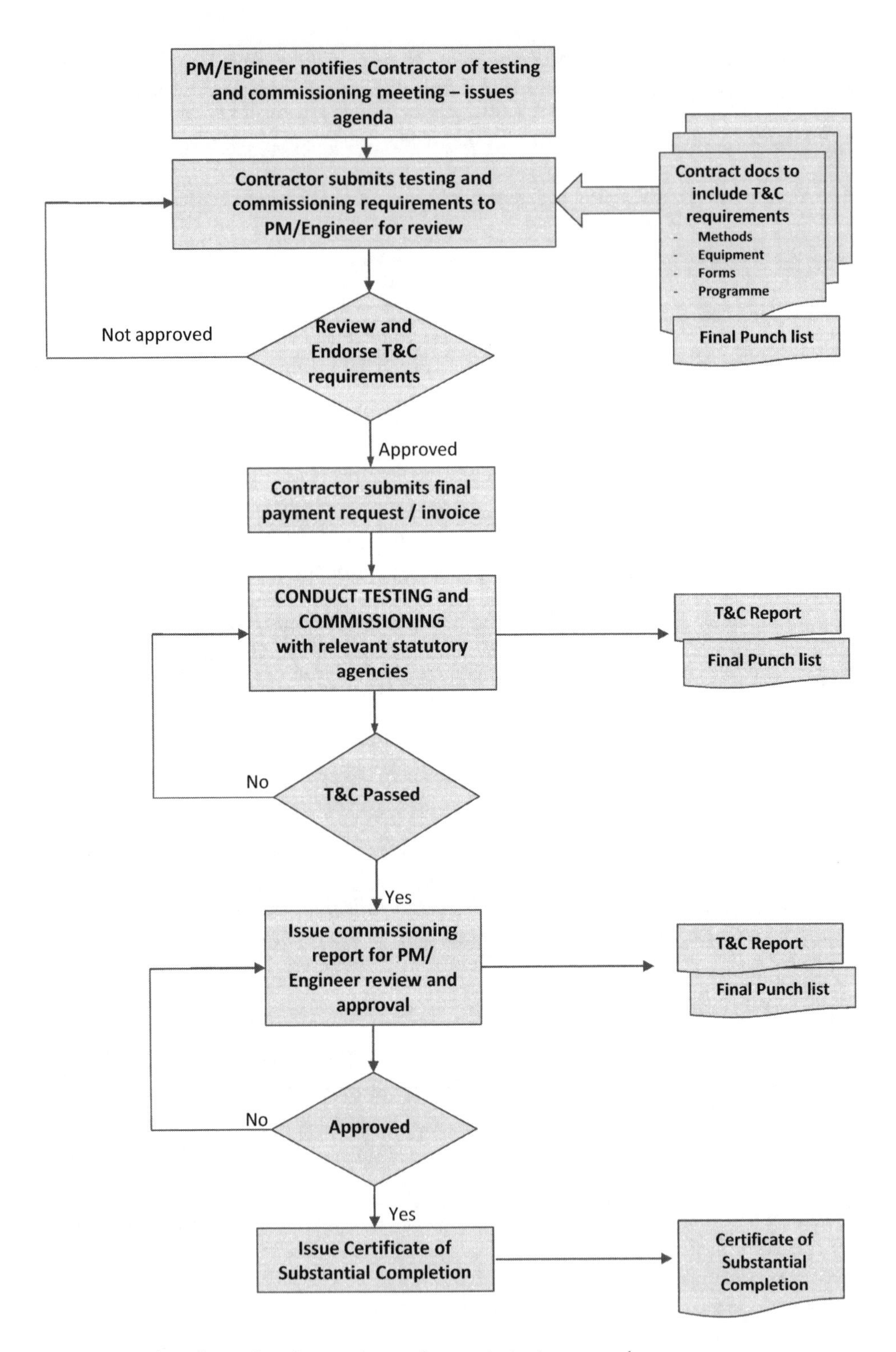

Figure 11.1 Flowchart of project testing and commissioning procedure.

been fully resolved and concluded by this stage. In other words, this project is now done and dusted, let's learn from it and move on.

The normal procedure is that, upon receiving the contractor's notification of completion of the corrective measures required in the final punch list / snag list:

a The contractor will submit for the employer's or PM / engineer's review and endorsement the Notice of Final Completion.
b The employer or PM / engineer, supported by the supervision consultant (SC) will review, verify and endorse the Notice of Final Completion, and issue the Certificate of Final Completion.
c Upon such approval the contractor will submit the final payment request.
d The PM / engineer will review and endorse the final payment request after necessary adjustments, if any, due to penalties, liquidated damages and other deductions.
e The PM / engineer will issue the Final Payment Certificate, and ensure document completion and dissemination of lessons learned, and prepare the Project Close-out Report.
f The employer will authorise and release the final payment.
g The project databases will be closed, lessons learned documented and stored for future reference, and any required project post-implementation review will be conducted.

See figure 11.2 for a flowchart giving the generic procedure for contract close-out. Clearly, individual employers and projects may vary this procedure to comply with their own individual circumstances and corporate procedures.

Table 11.2 also provides a useful list of inputs and outputs for the contract close-out stage of a project.

11.3.1 Contract Close-out Report

As mentioned above, both the employer and contractor will require closure on this project and therefore, for their own audit purposes, they will usually require a Close-out Report. This report should provide for the following:

1 Date of formal end of the contract
2 Confirmation that all penalties have been applied
3 Confirmation that all claims or disputes have been resolved
4 A record of the performance of the contractor
5 Confirmation that all performance bonds are correctly released as appropriate
6 Any free issue / employer supplied materials are reconciled
7 Final account is agreed with the contractor
8 Final payment has been made.

If the employer has a major programme of construction and development, they will also need to formally record the close of the contract in appropriate internal company IT systems as well as inform any internal audit departments if that is a company requirement.

11.3.2 Contract close-out documentation

The following contract documentation is normally required to be included in close-out procedures to ensure that all contract administration issues have been addressed and satisfactorily completed.

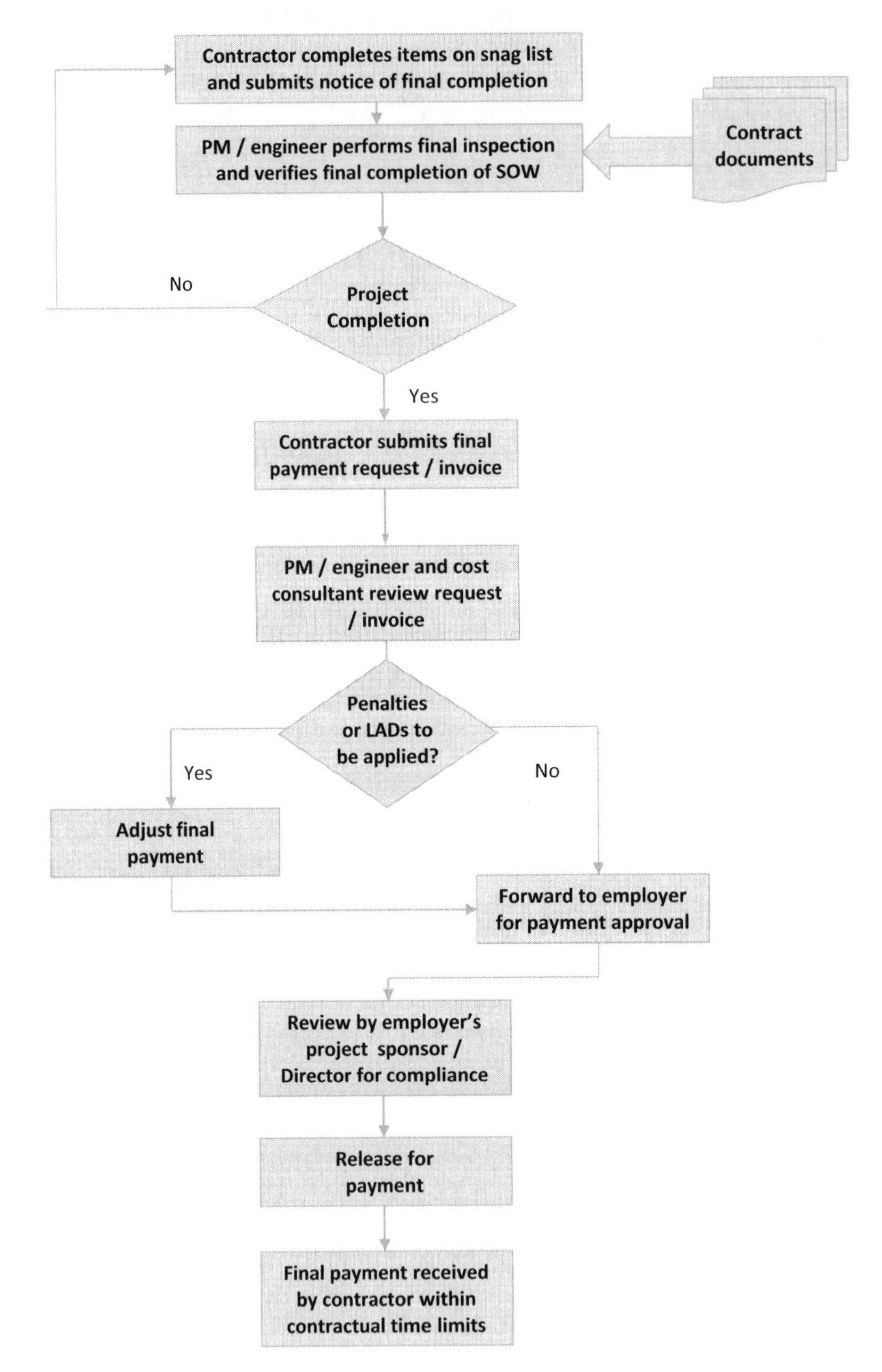

Figure 11.2 Flowchart of project contract close-out procedure.

Table 11.2 Inputs and outputs for contract close-out stage

Inputs	*Outputs*
Final punch list / snagging list	Certificate of Final Completion
Certificate of Substantial Completion	Project Close-out Report
Certificate of Making Good	Lessons learned
Original contract documents	Final Payment Certificate

Substantial Completion Certificate

Copy of the certificate issued at the time of substantial completion with snagging / punch list issued to the contractor listing the outstanding work required to be completed during the defects correction period.

Certificate of Making Good

Copy of the certificate issued following successful rectification of all defects.

Completion Certificate

If required under the particular form of contract used on the project, a Completion Certificate is normally issued upon achievement of the following:

a completion of all work including list of outstanding work;
b completion of demobilisation (if necessary);
c handover of all required manuals, documentation and as built information.

The Contract Completion Certificate therefore confirms:

a the end of the site activities and commencement of any manufacturer's warranty period;
b the release / adjustment of Bank / Performance Bond, as applicable.

Where a repair or replacement is required under a manufacturer's warranty, the warranty period should be further extended and this should be confirmed to all parties.

Statement of Final Account

The Statement of Final Account must be included with the close-out documents which identifies:

a original contract award / value
b summary of contract variations and agreed values
c summary of penalty / incentive payments
d final account value.

Contractor's written statement

Stating that all financial and other issues under the contract are settled and no invoices or claims are outstanding.

Contract Performance Report

To record the execution of the contract, including technical and commercial issues and lessons learned for future projects.

The content of a Project Report / Contract Performance Report may include:

- scope of work completed plus additions due to variations and employer instructions;
- final cost compared to original lump sum / target cost;
- balance of funds available (internal employer only);
- completed within contract period or extensions of time granted;
- major risks that have occurred, or passed without occurring;
- major remaining identified risks (passed on to end-user);
- HSE performance – any reportable incidents or lost time incidents (LTIs);
- results of variance analysis: e.g. schedule variance and cost variance;
- performance indexes: e.g. schedule performance index and cost performance index;
- summary of major approved change requests during the reporting period etc.;
- lessons learned.

11.4 Summary and tutorial questions

11.4.1 Summary

At this stage in the project lifecycle the building is finished, and all that is left is a period of rectifying any minor snags so that the final completion can be certified and everybody goes home. Although testing and commissioning is included in this chapter, in reality all appropriate tests will have been carried out before issuing the Certificate of Substantial Completion, as the building will not be fully usable unless and until all the systems are in full working order. If there are any minor snags to these systems, these snags will be rectified during the defects correction period and further tests carried out if appropriate.

During the defects correction period, the employer cannot issue any further instructions or variations, except as related to snags. If the employer requires any further work to be carried out, this will be subject to a separate contract or Works Order and the contractor is not bound by the valuation procedures in the main contract – they are free to negotiate separate terms.

When all contractual obligations have been fulfilled, including rectification of all snags and defects, the final completion and close-out procedures will follow, representing the final completion of the project. Hopefully, everything will have gone smoothly and all parties will be satisfied with the outcome – if not, the procedures in Part D of this book may come in useful.

11.4.2 Tutorial questions

1. Give examples of major defects and minor defects. How should each be dealt with under normal contract rules?
2. What is a latent defect? How may these be dealt with if the defect does not manifest itself until after final completion? (NB – this question is difficult.)
3. What is the purpose of a snagging list and when should it be created?
4. Outline the rights and responsibilities of all parties in a construction contract during the defects correction period.
5. Is ‘final completion’ actually final?
6. Outline the required contents of a Contract Close-out Report.

Part D
Claims and disputes

Disputes between the parties to a contract or between the various parties engaged on a project (who may have no contractual relationship with each other, e.g. the architect and the contractor) can arise at any stage in the project, so it is salutary to ask the question 'When are disputes generated?' This is actually quite a scary question as perhaps the best place to start looking at the origins of disputes is at pre-tender stage, when everybody is still bright eyed, bushy tailed and enthusiastic. The pressure on the various consultants to get a project off the ground with insufficient time for proper pre-tender design and planning is a rich source of eventual disputes during the construction stage, as corners may have been cut leading to poor design coordination, ambiguities in the design and other contract documents and therefore potential delays.

Disputes also often arise because someone failed to do something (either failed to do it at all, or failed to do it properly and thoroughly) that they ought to have done, or alternatively did something that they ought not to have done. At pre-tender stage, these failures may manifest themselves in the quality and standard of the various contract documents, such as the drawings, specifications and pricing document. Ambiguities and errors always create fertile grounds for disputes at a later stage. Therefore, following the old maxim that time spent on preparation is never wasted, a little more care and time in the design stage may actually and paradoxically shorten the construction stage by reducing the number of potential delays. However, it is a very mature and sophisticated client who will understand this argument at the design stage.

During the course of the construction stage, contractors have been known to exploit errors and ambiguities in the project documents and the employer's consultants (who will probably have been responsible for the poor documentation in the first place) often make matters worse by refusing to recognise the problem or their own culpability and dealing with it promptly and effectively.

Earlier standard forms of contract (i.e. those issued prior to 1998) did not cater very well for contractor's claims. If the contractor did have grounds for a claim of any sort, in most cases the claim was submitted at the end of the contract and ignored in the hope that it would go away. Designers generally had more time in which to complete the design prior to tender and all necessary details were usually provided at tender stage. If the contractor had done a good job, an amicable settlement was usually possible and everybody slapped each other on the back and moved on to the next project. Modern forms of contract, whilst having more sensible procedures for managing claims (such as the requirements for notices and early warnings etc.), can breed disputes because of the greater detail of the clauses and unfair risk allocation, especially if the standard form of contract has been amended by one of the parties. For example, if the contract requires the contractor to give notice within 14 days of every delay and cause for a claim, there are at least two possible scenarios:

1 The contractor may not be able to strictly comply with all of the provisions and therefore their application (claim) is rejected on procedural grounds.
2 The contractor will give notice for every single potential claim (no matter how trivial) thereby creating bad feeling throughout the project.

Clearly, neither of the above scenarios will be beneficial for the project, but in the modern construction industry, drafters of contracts positively invite the second possibility. It is very often difficult to distinguish between the competent contractor who is merely following the contract to the letter, and the claims conscious contractor who will claim for everything whether or not there are proper contractual grounds.

It is possibly a sad reflection on the industry that a book of this nature needs to have four chapters on claims, disputes and formal / informal methods of dispute resolution.

12 Legal basis of claims and claim management

12.1 The legal basis covering contractual claims

12.1.1 The contract

During a construction project, many problems can be encountered which were never foreseen during the original contract negotiations between the parties. A claim is the articulation of that problem in the form of a Statement of Entitlement by the contractor to additional time or cost. A contractor may have a claim against the employer or a sub-contractor may have a claim against the contractor for a variation or a delay. Conversely, the employer may have a counterclaim against the contractor for another delay or defect which, in turn, the contractor may wish to pass on to the sub-contractor. All in all, it can get to be a very messy and protracted business.

Claims arising under a construction contract generally fall into one or more of the following categories:

- an increase of work or variations / changes to the contract scope of work;
- an extension of time, together with associated costs (also termed loss and expense) as a result of delays and / or disruption to the contractor's works;
- claims on a *quantum meruit* basis, i.e. a claim for the works done or materials supplied on the basis of 'for what it's worth', i.e. a reasonable price for the work if there is no procedure for valuation in the contract;
- liquidated damages claimed by the employer or main contractor for the late delivery of the works by the contractor or sub-contractor; and
- a claim for defects or poor workmanship.

Claims are therefore simply Statement of Entitlements and are effectively a formal contractual application for an extension of time and/or interim payment / deduction which, being provided for under the terms of the contract compensation procedures, provides a quicker and cheaper method of resolving disputes than any of the formal dispute resolution procedures discussed in the following chapters. They should be seen as integral to the administration of a construction contract as they provide a certain flexibility to deal with any problems or issues which may be encountered. It should be noted that these are claims for entitlement to money or time under the terms of the contract between the parties; by definition, the contract cannot deal with claims under tort (e.g. negligence) or other extra-contractual issues which may arise.

12.1.2 Breach of contract

As an alternative to a claim specifically relating to a clause of the contract, a party may be able to show that any loss or expense incurred is recoverable as damages following a breach of contract by the other party. As we have seen previously, damages are the normal remedy for breach of contract and should be calculated simply to compensate for the loss suffered. Such a claim will clearly depend on the specific obligations contained within the contract, the facts of the circumstances and the impact of such a breach. In addition to a breach of explicit clauses, the injured party may also be able to rely on certain implied terms which will depend on the jurisdiction of the construction contract.

These implied terms imposed on the other party (or their representatives, such as the PM / engineer) may include the following:

- warranty of the plans and specifications
- duty to provide access to the site
- duty to provide adequate supervision
- duty to act in good faith and fair dealings (especially in US law)
- duty to provide a safe workplace
- duty not to hinder or delay the progress of the works.

If such a term is found to be implied, any act of hindrance by the employer or PM / engineer would, quite possibly, give rise to a claim for breach of contract; provided the loss is not too remote, that the breach is directly attributable to the employer or their designated representatives, and the contractor can show that the loss or expense is as a result of that breach. In such a case, the contractor may be able to recover their losses using the ordinary principles of damages quantification, provided it can substantiate the loss with the required level of proof. Proving the contractor's loss clearly requires full records to be kept during the execution of the works – so proper contract administration requires:

RECORDS RECORDS RECORDS

The need to keep full and complete records of what has happened on the project cannot be overestimated.

Claims for loss are meant to put the claimant back into the position it would have been in had the event not taken place; they are not meant to provide a windfall or to turn existing losses into a profit. The losses should also be reasonably foreseeable, and must be considered as naturally arising as a result of the course of events from the breach. Losses may also be recoverable when it is reasonable to suppose that they were in the contemplation of the parties; in this sense such losses may also include a loss of bargain, wasted expenditure or restitution.

12.1.3 Tortious claims

Most construction disputes and legal cases deal with contract law, not the law of tort. The difference between contract law and tort is mainly that contract law is based on enforceable written agreements, whereas tort is based on a 'personal wrong' committed by someone against someone else (the word 'tort' is derived from the Latin for 'wrong / injustice').

There are many different types of tort, but the type which is most appropriate in the construction industry and generates the vast majority of claims is that of negligence. In many cases, if a claim in contract against the contractor or sub-contractor is not possible, a plan B would be to pursue a claim for negligence against the designer, as it may be the design rather than the erection which is defective.

Negligence occurs when a party fails to demonstrate the kind of reasonable care that is expected of a qualified practitioner and an injury results from the action or inaction. There are five elements necessary to prove a negligence case:

- defendant owed a duty of reasonable care;
- defendant did not behave in a reasonable manner to demonstrate care;
- plaintiff suffered an injury as a result of the defendant's actions or inactions;
- the injury caused actual damage;
- defendant's actions or inactions were the cause of injury.

The modern law relating to negligence in the UK was set down in 1978 by a major legal case (*Anns* v. *Merton* if you really want to know) but was turned completely on its head in 1988 by another major case (*D & F Estates* v. *Church Commissioners*) which concluded that only physical loss can be recovered through an action in tort – economic loss cannot be recovered. This particular bombshell was the catalyst for the development of collateral warranties whereby an employer or end-user would be able to take action under a contract, whereas in the past they may have relied on a tortious liability which is now not available. This is particularly relevant where an employer has a complaint against a sub-contractor or supplier as they do not have a direct contractual relationship (see section 7.2 and figure 7.1 in Chapter 7).

12.1.4 In restitution / unjust enrichment

The purpose of the law of restitution is to redress any *unjust enrichment* which one party to a contract may have gained at the expense of the other. The law works by attempting to reverse that unjust enrichment and giving any relevant benefit back to the claimant to put them in the position they were in before the unjust enrichment. Circumstances where this may arise could be where a contractor performs some enabling works at the request of the employer (possibly via a Letter of Intent or separate Works Order), which is carried out in anticipation of a contract but that contract never materialises.

So, when can a contractor claim for unjust enrichment? Firstly, there must not be a contract currently in place between the claimant and respondent – otherwise the claimant would have to take action under that contract. Providing this is the case, a claim in restitution must then establish three things:

- that the defendant has received some sort of benefit. (i.e. they have been 'enriched');
- that the enrichment was unjust (i.e. there was no reciprocal benefit);
- that the enrichment was at the expense of the claimant.

Following this, the claimant must also provide details of the following:

> *Nature of the benefit received*: If there is some evidence that the responding party received incontrovertible benefit, for example an immediate financial gain or a saving, then it is more likely that enrichment can be established.

There was no understanding that the services were given freely? If the defendant accepted the services when offered to them and knew they were not intended to be given freely, the court is more likely to impose restitution. Clearly, most companies are in the business of making profit, so it is highly unlikely that they would perform work for free, especially if the work was substantial in nature.

Unconscionable behaviour: That the defendant deliberately declined, or refused, to make payment.

Assumption of risk: If the claimant has taken the risk that they would not be paid in the event that a contract was not concluded then it is unlikely that restitution would be imposed.

Preliminary work: If the work carried out by the claimant was simply work preliminary to a contract, for example where the claimant carries out works that will put it in the position of being able to win or perform work on a future contract, then there is no clear direct benefit to the defendant therefore restitution may not apply.

Therefore, although this is clearly a difficult area and, as with any legal procedure, there are no guarantees of success, restitution should not be forgotten as a potential route to recovery should there be no formal contract in place. Internationally, the concept of unjust enrichment is also an important aspect of Shari'a or Islamic law. Therefore, for projects taking place in, say, the Middle East, the civil code covering the project will include for this legal principle.

12.2 Claim management

12.2.1 Strategy

A contractor's claim should provide sufficient information to the employer to be able to persuade them that there is indeed an additional entitlement to either time or costs under the contract. Most construction claims require a Statement of the Facts, a convincing technical and legal interpretation of the particular contract terms, together with an explanation of the engineering or logistical consequences set out in terms of time and / or cost. The extent of analysis and explanation will clearly depend on the scale of the issue being considered, the size and nature of the project as well as the experience of the parties.

Claims are not the same as disputes and do not necessarily create disputes, if the employer or PM / engineer accepts that there is a valid argument being put forward. However, in the unfortunate but all too real world of confrontation between us and them, the possibility that the parties will not agree must be seriously considered and it is highly advisable for a contractor, in order to save a great deal of time and effort in the future, to prepare a structured approach to claims at the outset. Firstly, each forum or method of dispute resolution must be considered and the management of each party's expectations within that forum will be key to establishing a structured and effective approach in the resolution of a claim. The amount and detail of proof and evidence needed to convince a party will rise in direct proportion to the other side's opposition to settling the claim at the level predicted. The management of claims requires a speedy establishment of a realistic and sensible level of expectation on both sides – i.e. contractor and employer. To do this, it is important to understand

and appreciate the other side's position and argument, therefore effective and willing communication is extremely important.

In order to manage and resolve a claim efficiently and effectively, it is paramount that both the legal and technical investigations are carried out thoroughly and rigorously using best practice techniques. These two elements are clearly interrelated and dependent on each other but they also require wholly different skill sets. Therefore, to manage this process, some flexibility coupled with an effective and detailed strategy will be required.

To prepare such a strategy plan, an initial analysis will be needed, on small projects this may involve a meeting with the engineer or project manager whereas, on larger or more complicated projects, this may require a series of detailed and intensive meetings with someone legally qualified who is experienced in construction claims management.

The strategy plan, usually in the form of a network programme, will be the device which will be used to manage resources and report progress. The use of efficient claims management processes and the accurate forecasting of losses to be incurred, on both sides, should be set against the gains to be made at each forum and the chances of success at each level. It may be possible to construct decision tree models to analyse each possible outcome. In complex construction claims, this in itself will require considerable investment of time and money.

12.2.2 Evidence

Evidence is a very complex area in law and as stated previously, this is not a legal textbook; however, the claim document produced by the contractor must follow the basic principles to ensure that it is not rejected at the first hurdle. Evidence is the proof of the facts that have been stated and taking the general rules, should be sufficient, authentic, relevant and current.

The management of evidence requires the efficient and simple tracing of documents in order to retrieve relevant facts at any stage of the process. There are a variety of electronic systems available but all now involve some form of electronic document management. This includes the creation of electronic copies of the documents which have been relied upon, with some form of indexing by document type, date and reference number. As the process progresses, the index expands with subject references and reference numbers. Storage, using hard data storage or cloud facilities, will provide access for all team members, dependent on geographical location.

12.2.3 Analysis

The analysis of delay and disruption must be made on a factual basis with an emphasis on the engineering and logistics of the project. The selection of the proper analysis method will consider a variety of factors including the information available, the time of the analysis, capabilities of the methodology and time, funds and effort allocated to the process. The management of this evidence in relation to these aspects requires the scheduling of activities, dates (start and finish), the construction logic, together with any restraint activities. A schedule will also provide an explanation as to why particular activities cannot start until a previous activity has finished: an explanation of any construction restraints on the start of activities, and an explanation of the main operations which dictate the periods of activity. Document references are scheduled alongside each activity and the details of the schedule are incorporated into the as-built programme.

12.2.4 Settlement

It is essential to develop an overall strategy in the management of a claim which centres around the aim of achieving early agreement to entitlement at the optimum commercial value. Firstly, it is important to determine the relative importance of some factors that influence the accomplishment of any commercially efficient negotiations; secondly, it is equally important to put a value on the so-called BATNA (Best Alternative to a Negotiated Agreement).

BATNAs are considered as a basic strategy negotiation tool: the procedure followed may enable the parties to come to a satisfactory agreement leading to the achievement of commercially efficient negotiations. This can be achieved by the targeting of three important aspects:

- the importance of obtaining precise information;
- the need to come to efficient results;
- the need to extract operative implications which are necessary to accommodate flexible negotiations.

To implement this strategy it will be necessary to identify a likely settlement range which takes account of the strengths and weaknesses of the claim, the attitudes of the parties and the identifiable pressures each party is facing. As the claim and negotiations develop, the pressures and settlement range is likely to change.

During negotiations, it is important that the negotiators have the persuasive skills to present a robust and convincing case. They must be technically, legally and commercially competent and able to identify and deal with any new evidence which strengthens or weakens the case and accommodate any changes to the settlement range.

12.3 Components of a claim document

How a claim document is structured and presented by the contractor will often determine whether there is a quick and inexpensive resolution of the claim or whether it develops into a prolonged and expensive battle, possibly involving formal arbitration or litigation proceedings.

As soon as the contractor's management has determined that a particular claim should be pursued, comprehensive preparation and organisation is essential and must be promptly undertaken, especially if there are any time limits in the contract. The commercial department of the contractor should immediately assemble, organise and review the relevant facts, evidence and documents which have a bearing on the claim, while memories are still fresh and before facts, evidence and documents are forever lost or forgotten, or critical staff move on. Although early resolution through informal procedures is clearly the best strategy, it is prudent to always prepare claims with an eye on the possible end-game – i.e. resolving the claim in the formal setting of arbitration, litigation or another dispute resolution process. If early resolution is not achieved, complete preparation at this early stage provides a solid foundation for future formalities.

Claims and disputes involving construction projects tend to be technically complex and contain enormous amounts of facts and evidence. The initial claim document is a written synopsis of the claim that can be presented to the other party at the early stages of the dispute – providing, of course, that there is a 'dispute'. In some cases, a formal claim submission is required by the contract, therefore the contractor must follow the contract requirements including submission of notices and deadlines for providing supporting documentation. Whether a formal or informal process is followed, the immediate and primary goal of preparing and submitting a claim document is to bring about a prompt and satisfactory resolution of the claim through negotiation. Failing a satisfactory resolution of the claim, a well-prepared claim document provides a blueprint or plan for further formal claim management procedures.

KISS – Keep it simple, stupid

The key to effective claim preparation and presentation is keeping it simple. Don't try to be too clever with the working as simplicity promotes understanding, especially if the employer is a non-technical person. The process of preparing and presenting a claim document is an important step in developing the overall claim strategy, as it requires an easily followed logical progression from beginning to end. While the claim must be well supported by evidence, the claim document should explain the dispute in a simple but complete and comprehensive manner.

Tell the story

The claim document must tell the story of how and why what actually did happen was not what was planned. It should be interesting without being banal or shallow. There must be a clear and definite theme that is communicated and readily understood and which is the strongest argument supporting the argument.

Include an executive summary

As with all management reports, the executive summary is not an introduction; its purpose is to summarise the report and any conclusions. This will be followed by the factual narrative in the main body of the document. As stated above, construction claims can be technically complex and factually intensive, therefore it is often helpful to include an executive summary before immersing the reader in a long and complicated narrative of facts and data.

The main body of the claim – factual narrative

The factual narrative will focus on the intended outcome but should not be expressed in an overly argumentative or aggressive manner – the facts should speak for themselves (the legal principle *res ipsa loquitur* means exactly this, although mainly used in the law of tort). The writing style should be clear and precise but not dry – it is not the project's technical specifications. It is, after all, a story, not simply a recital of a string of facts. A narrative that is comprehensive and logically organised will provide a good resource that can be used in later negotiations if necessary.

Emphasis on the strongest points, with key facts

When multiple and unrelated claims are presented in one claim document – i.e. a 'global' claim, the document must be structured to emphasise the strongest claim first with a focus on the key facts supporting the claim. Presenting and arguing every fact one after the other may simply overwhelm and confuse the reader, with the result being that the claim document is rejected and back to the drawing board.

Provide the contractual and legal basis clearly and effectively

Depending on how the claim is structured, there will need to be a written discussion of the contractual basis supporting the claim, plus any applicable legal principles that support and illustrate the basis of the argument, if necessary. It is always beneficial to seek assistance from an experienced construction claims consultant, or at worst, a lawyer to develop the legal arguments and to ensure that the factual narrative is presented in a manner consistent with the applicable legal principles and specific contract provisions. The need for, or extent of, a legal discussion generally is geared to the expertise or experience of the ultimate decision maker for the employer and may act as a crucial element in their understanding of construction law and their liabilities, all of which may help in early settlement of the claim.

Provide detailed and accurate cost and pricing data

Pricing the claim together with supporting information on damage calculations is just as important as establishing contractual and legal liability for your claim. This is often referred to as the 'quantum' of a claim – i.e. the amount that is being asked for. Providing inaccurate or defective quantity, cost and pricing data may subject the claimant to serious allegations of false claims, which then exposes the claimant to possible significant penalties and financial liability. The claim document should specify the actual financial amount claimed and must also include a fairly detailed cost analysis and breakdown of the claimed damages. If not too voluminous, an indexed appendix with all relevant supporting documentation should also be provided, although this may not be fully required until the claim reaches a more advanced stage. Any third party sources that have been relied on to support the cost and pricing data must be referenced and acknowledged.

Highlight the most persuasive documentary evidence

The most crucial and important documents should be quoted in the body of the factual narrative; those documents that do not merit full incorporation into the text, but which also support the claim, can be included in an indexed appendix cross-referenced with the factual narrative. This clearly simplifies the structure and readability of the main claim document.

Include Expert Reports and evidence from similar claims

Charts, graphs, drawings and especially photographs are very helpful in demonstrating points made in the factual narrative and should be incorporated into the claim document wherever possible. Similarly, relevant reports by acknowledged experts may be included as attachments or exhibits, with appropriate references to and quotes from the reports in the main narrative. Clearly, commissioning a third party to produce an Expert Report will add further costs to the exercise which may or may not be recoverable, and it may also be better to keep the expert opinion until a later stage in the process.

As mentioned above, the advice and guidance of a construction claims consultant or expert is worthwhile at the earliest stages of claim preparation unless the contractor has competent in-house expertise. A seasoned consultant can help manage the dispute so that it can be settled well before any expensive formal procedures such as arbitration and litigation. If the dispute does go to a formal procedure, the claims consultant will be able to prepare relevant documentation and liaise with the legal profession as required. All supporting documentation and evidence must also be structured in formats required by the rules of the appropriate procedure.

12.3 Summary and tutorial questions

12.3.1 Summary

Contractual claims are named as such for a reason – they need to be based on the terms and conditions of the contract between the employer and contractor. The term 'claim' has had a bad press and conjures up images of contractors trying to extract more payments from the contract than they are entitled to. In the majority of cases this is unfair and as it is a very expensive business for a contractor to prepare and submit a claim, they should only do so when there is a good contractual justification for the additional payment. Outside of the actual contract between the parties, the general law may also be used for ex-contractual claims with the tort of negligence and the concept of unjust enrichment being useful legal principles for a contractor to show an entitlement to payment which may have been previously refused.

The actual claim document submitted by the contractor will be structured in accordance with the details of the individual circumstances, but in general, will need to include:

- a Executive Summary, summarising the issues, contract conditions, aggregate quantum and total amount requested as entitlement;
- b Factual narrative giving the history of events;
- c Contractual terms and conditions which have been relied upon;
- d Quantum of claim (in terms of m^3 / m^2 etc. of appropriate Bill of Quantities items with cost data);
- e Programme / schedule effects (i.e. required extensions of time – see also Chapter 13);
- f Expert Reports, to back up the case for entitlement;
- g Conclusions.

From a contractor's point of view, they are not submitting a 'claim' but an application for 'entitlement to additional payment' under the terms and conditions of the contract. There are therefore three major rules to consider before submitting the application:

- read the contract,
- read the contract again,
- re-read the contract,

to ensure the issue covered in the application has a robust contractual justification.

12.3.2 Tutorial questions

1. Define the term *quantum meruit*.
2. Give potential examples of unjust enrichment as a contractor's claim for entitlement to payment.
3. Outline the difference between a claim and a dispute.
4. Why is it essential to keep full contemporaneous records of construction activity?
5. Discuss the structure of a claim document and when this should be presented by the contractor.
6. What types of evidence should be included in a claim document?

13 Claims for extension of time, delay and disruption

13.1 Purpose of extension of time claims

As discussed in Chapter 10, the majority of construction contracts include an express provision for the completion of the works by a certain date; this date may be calculated as a period of time after the commencement date (the 'contract period') or it may be inserted in as an actual calendar date in the appendix to the contract. If there is no date for completion contained in the contract – i.e. 'time at large' (see section 13.3.1 below), a term will be implied requiring the contractor to complete the works in a reasonable amount of time. The contract will also usually require the contractor to provide a programme of works, for approval or acceptance by the PM / engineer, in order to demonstrate how the works will be completed, and in what sequence.

Chapter 10 also states that construction contracts will also typically include an express provision which facilitates an adjustment to the completion date in certain circumstances. These extension of time clauses are inserted into the contract to cater for the occurrence of specific risks or a breach of contract by the employer and will push the completion date forwards in time if the contractor can prove entitlement. The procedures contained in the particular contract normally require very strict compliance before the contractor is entitled to, or awarded, the extension of time. Whilst an extension of time clause does facilitate a change to the completion date, its primary purpose is to protect the employer's right to charge liquidated damages for any late completion by the contractor.

13.1.1 Definitions of delay and disruption

The terms 'delay' and 'disruption' are sometimes (incorrectly) used interchangeably in the construction industry. Delays occur when the contractor is prevented from continuing with the works as originally planned; i.e. the contractor is late in delivering the works. There are therefore two ways to look at the word 'delay' in construction contracts. The first, from the contractor's viewpoint, would be a delay to their own progress which may affect both critical activities and activities in float (see section 10.3.3 for a definition of float in this context). These may or may not lead to time and monetary compensation for the contractor once it is established who has caused the delay and who bears the risk for those causes. The second viewpoint, from the employer's perspective, would be delay to completion. These kinds of delay are usually caused by events which affect the critical path and, where those events are attributable to the contractor, lead to the application of liquidated damages under the contract. Delay is relatively easy to measure insofar as the actual progress of the works is compared against the anticipated progress in the contract programme or baseline schedule.

Disruption is a term used to describe the effects on an activity which is affected by a change, prevention or hindrance to its expected progress. Whilst it is also a comparative term; i.e. in order to prove its effect it must be compared against a benchmark, it is not the same as delay, and must be considered in the context of that particular activity's productivity. Whilst disruption may cause delay to an activity, it conversely may also result in an activity being accelerated. In either case, whether delay or acceleration, it will result in additional costs as a result of that change in productivity with additional expenditure incurred and an effect on the contractor's method and sequence of working.

13.1.2 Notices of Claims as condition precedent to entitlement

Most procedural requirements contained in construction standard forms of contract require the contractor to provide notice when a delay or disruption first occurs. The intention of this notice is twofold. The first is considered as a management aid insofar as it provides the contract administrator or PM / engineer with the advanced notice necessary to take decisions or action to mitigate the problems being encountered. The second important factor is to facilitate entitlement for the contractor to claim payment for the effects of the delay / disruption. In this sense, and to encourage the contractor to provide timely notice so that the contract administrator can take such evasive or proactive action as necessary, the notice by the contractor is normally made a condition precedent to any entitlement.

A ***condition precedent*** is a crucial concept in contract administration, and is an event that is required before something else can or will occur. In many of the standard forms of contract, a notice by the contractor of a delay must be given before an extension of time can be issued. FIDIC Clause 20.1 gives very specific wording for these circumstances.

In order for the notice to be a condition precedent, there must a statement in the clause which provides the precise time when the notice must be served. There must also be express and clear language that failure to provide the notice within that time will result in a loss of entitlement. Where a notice is considered as a condition precedent there must be certain formalities which must be complied with so that the notice constitutes what is considered as 'proper notice'. These formalities include the form the notice takes, the details to be provided and the supporting information required.

Interestingly, in many legal jurisdictions where there is a concept of 'good faith' in all business dealings, this concept can, and has been, used as a defence by the contractor when such notices have not been served. Similarly, the concept of 'unjust enrichment' (see section 12.1.4) may be used by the contractor if the employer has rejected a fair claim for entitlement purely on a technicality. However, for a contractor to rely on such arguments is very dangerous and all contractors are well advised to keep full and comprehensive contemporary records, which can be used in future claims if necessary. In any dispute resolution procedure, the contractor's contemporary records form the basis of all evidence.

13.2 Completion of the scope of works

The contractor's primary obligation under the contract is clearly to complete the scope of works detailed within the contract documents. This is usually defined within a general statement of the employer's requirements, relevant drawings, specifications and standards as to how the work will be completed, together with such ancillary requirements (such as health and safety, environmental etc.) as necessary. Where no specification is provided, it will be implied into the contract that the work done will be to a reasonable standard of care and skill, using materials that are reasonably suitable for their purpose. Where specifications are provided; they usually address all manner of sundry matters including the method and form of working and any relevant tolerances.

13.2.1 Contractor required to mitigate any delays

Most standard form contracts require the contractor to take all appropriate steps to mitigate the effects of any delays. However, the extent of this obligation differs and, depending on the terms contained in the standard form, the PM / engineer will have a number of different ways of considering the contractor's actions in such circumstances. Therefore, failure to comply with an obligation to mitigate delays may have the effect of reducing or even extinguishing any entitlement to an extension of time entirely.

13.2.2 Effect of increasing and decreasing scope of work

As we have seen in Chapter 6, variations arise when there is either an increase or decrease in the scope of the work to be completed by the contractor. A need for a change to the scope of work may arise because a change is required in the design or that certain items were not included in the original contract, or even that the works contracted for cannot be carried out at all. Where such a variation is required it will be issued under the terms of the contract, which means that the contract stays the same but the scope changes. Many standard forms of contract include a clause which provides for variations to the scope of works and a procedure for the parties to follow in incorporating and embedding the change. These provisions are helpful insofar as that they preserve the contract whilst allowing the changes to the scope of work. The absence of such a clause may possibly entitle the contractor to refuse to carry out additional work; therefore these clauses allow the employer the flexibility to make any necessary changes and also entitle the contractor to extra payment and / or time.

We have also seen in Chapter 10 that the employer is not entitled to issue Variation Orders after the issue of the Practical / Substantial Completion Certificate. In many regions of the world, employers still request the contractor to 'do this' or 'do that', well after the issue of the Certificate of Substantial Completion. Clearly, a contractor will not want to refuse an employer's request to both maintain good relations and also make a little more money, but these requests should not be classed a variations or instructions under the contract and should be valued as completely separate contracts in their own right with cost and time being negotiated separately. They are often termed 'Works Orders' to differentiate them from Variation Orders. As they are separate contracts, the PM / engineer may also not be responsible for supervision, quality control etc. so care must be taken with their own professional liability.

13.2.3 Acceleration of the works

Where time has been lost on a project, it may be necessary to recover that time by instructing the contractor to accelerate the programme of works. Acceleration therefore occurs when the time for the works, or a portion of the works, is shorter than originally envisaged in the original programme. Many standard forms permit the PM / engineer to instruct the contractor to change the rate of progress and any payment of additional costs related to this instruction will be authorised as a variation and will depend on the circumstances which required the original instruction. Acceleration clearly has an impact on resource productivity by way of overtime and / or additional shifts, trade crowding or increased manpower. As the contractor will already have deployed the minimum number of workers required to perform the task within the allocated time frame, any increase that results in a higher rate of progress will have a higher unit cost. If the contractor is not responsible for the initial delay (termed a 'culpable delay'), as stated above, it will be necessary to compensate them through the variation procedures contained in the contract.

Acceleration or recovery?

However, and this is a big 'however', acceleration should be specifically instructed by the employer, which will be for their own purposes – i.e. to get the facility finished sooner than they had originally anticipated. If it is established during normal project supervision that the contractor is unlikely to finish by the contract completion date, the PM / engineer can request that a 'recovery programme' is put into place so that the contractor will finish by the scheduled completion date. This is clearly not an instruction to accelerate, but to get back on to the original programme. A contractor, possibly with a view to claiming additional payments, may misconstrue this as an instruction to accelerate, therefore the wording of any letters or notifications to recover lost time must be very carefully drafted.

13.3 The project programme / schedule

The initial project programme (i.e. the baseline schedule, BLS) may or may not be a contract document. In some parts of the world it is common for the BLS to be a legally binding part of the contract, but many of the major standard forms of contract (JCT, NEC and FIDIC) do not anticipate that it is a legally binding contract document. There are major advantages and disadvantages of making the initial programme a legally binding document, and these are fully discussed in more specialist textbooks, but overall, the industry appears to settle on the recommendation that it should not be a formal contract document.

13.3.1 Time at large

This is a concept derived from English law and may not be recognised in other jurisdictions. If the time for the completion of the works is described as being 'at large' (see section 10.4.1), the contractor does not have an inordinate amount of time to complete the works, but must complete within a reasonable time. What actually constitutes a reasonable time is a question of fact where all of the relevant circumstances of the project will be taken into account. Under normal circumstances, time is made at large in two scenarios:

a where there has been no stipulation in the terms of the contract regarding when performance is to take place or be completed;
b where time in the contract programme has been fixed but subsequent events, such as a breach of contract by the employer, or possibly *force majeure*, render it impossible to complete by that date and there has been no corresponding award of an extension of time from the employer.

13.3.2 Prolongation of the programme

As discussed above and at length in Chapter 10, under most forms of construction contract, the PM / engineer has the power to extend the time for completion. Assessments may be required at specific stages of the works or the PM / engineer may extend the time for completion once the contractor has followed the appropriate contractual extension of time process (including any notice and substantiation provisions); they may also agree the impact of any qualifying events, as provided for in the contract.

As part of that assessment, the PM / engineer will be expected to take account of any failings by the contractor. The PM / engineer is usually responsible for the notification and assessment of contractor culpable delay, together with attendant notice provisions. Any contractor culpable delay will, presuming the delay impacts the completion date / milestone dates, enable the application of liquidated damages.

13.3.3 Concurrent delays

As we have also noted, construction contracts apportion risk between the parties and also stipulate whether certain unforeseen events are excusable and therefore compensable to the contractor. Employer risk events and any related delays are usually compensable but the extent of any compensation (in time or cost) may be limited by what is known as 'contractor culpable concurrent delay'.

Concurrent delay is described as a period of project overrun, which is caused by two or more causes of delay, each of which actually delays the works and one (or more) of them is attributable to the contractor. 'True' concurrency (where delays caused by both the employer and contractor occur at the exact same time) rarely occurs, but it is generally accepted that both events may overlap and contain a concurrent period. To add a little more complexity, as though we needed it, there are two types of concurrency – concurrent cause and concurrent effect. Any claim for concurrent delay would have to establish and deal with this cause and effect issue.

For concurrent delay to exist, certain key factors should be present:

a The delays should impact the time for completion (and not simply absorb float).
b The delays should be independent of each other and should not be caused (directly or indirectly) by the other party's delay. And
c The delays should run in parallel, impacting at the same time but not necessarily be identical.

Figure 13.1 gives a simplified graphical illustration of concurrent delays; of course, the actual period of delay will be made up of a considerable number of individual activity delays and will certainly not be as clean as indicated in this diagram.

In order to properly analyse concurrent delays, each delay to an operation or activity must be considered separately by assessing its impact on linked activities and the project as a whole. The periods of delay attributable to the contractor would then be deducted from the overall period of delay and would not normally be subject to an extension of time. This whole issue is fraught with arguments; for example, the delay caused by the employer may have had more of a 'dominant effect' than the delay caused by the contractor, so why should the contractor suffer? 'Such is life' is the normal response, which means that the contractor must have watertight and airtight contemporary records if they stand any chance at all of being awarded additional time and especially additional costs for delay and disruption to the works. This principle of 'dominant effect' is the main method of analysing concurrent delays under English law, but in many other jurisdictions the prolongation costs may be apportioned depending on the delays caused by the employer and by the contractor. This principle of 'apportionment' appears to be gaining some traction in the international construction industry and slowly superseding the 'dominant effect' principle.

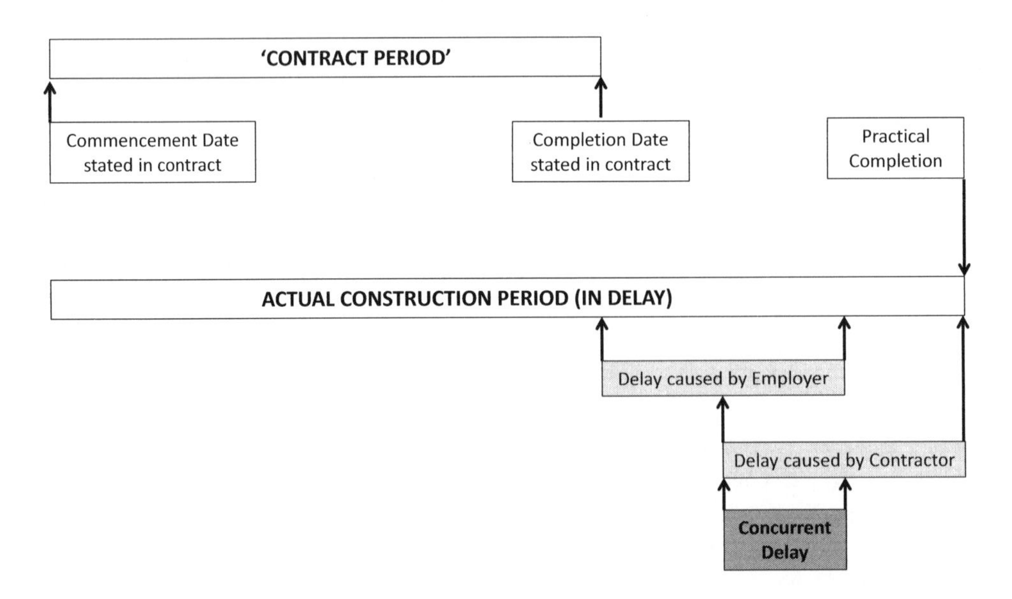

Figure 13.1 Concurrent delay.

Delay analysis is a massive area in construction planning and control and an introductory text such as this can only scratch the surface.

13.4 Employer's implied duties

As discussed in Chapter 1 and in common with all contracts, the employer, as one of the parties, must also perform its obligations under the contract. These obligations may be explicitly written into the contract or they may be implied in order to make it (and the contractor) work. Obvious undertakings include the supplying of information and the giving of possession of the site at the time the contract programme provides, whilst others may include a duty to act fairly towards the contractor or not to hinder the contractor's progress. Other duties also arise in terms of health and safety and other legislation related to the country of the project.

If the work is to take place within an existing facility, the employer will have further levels of responsibility in addition to those where the contractor's works are to be constructed on a new site. This responsibility stems from a need to control the existing operations whilst coordinating and cooperating with the construction workforce. Other than the supporting documents accompanying the Construction (Design and Management) Regulations (in the UK) there is little to guide an employer in how to discharge its share of responsibilities in these matters.

Matters that should be considered include the short-listing of potential contractors on the basis of their health and safety record, the provision of relevant information to the designer, arranging surveys to discover the extent of hazardous conditions, allowing sufficient time to carry out the work safely, the supervision and segregation of works from normal operations and not interfering with the contractor's operations.

Under the CDM Regulations (2015) in the UK, there are several distinct roles to be appointed by the employer – designer, principal designer, contractor, principal contractor etc. In all cases, the appointees must be competent and able to devote adequate resources to health and safety. The employer must ensure that sufficient resources will be allocated to enable the project to be carried out safely.

13.5 Delay analysis and evaluation

When undertaking delay evaluation, it is extremely important that the analyst understands both the construction process and the logic contained in the programme. It is also important to understand the limitations of the programme software and all the detailed project records and evidence available so that any analysis can be explained to the person to whom it is addressed – who will most likely be a non-technical person in the employer's organisation.

The records and evidence available to the analyst will determine the method to be used, which in turn will demonstrate any entitlement of additional time for the project completion. Many standard forms of contracts specify the form and content of the information to be recorded on a construction project and this, together with the original and updated programmes, should provide a sufficiently detailed and convincing explanation of the delay. In addition, it is also necessary to differentiate between delay to work in progress (prospective analysis) and a more forensic analysis after completion (retrospective analysis). The techniques described below show which they are more useful for.

According to Rider 1 of the SCL Delay and Disruption Protocol, there are six main methods of delay analysis:

a *Impacted as-planned*, which requires the identification of the baseline programme and the modelling of the event in question, with the use of planning software.
b *Time impact analysis* is a highly detailed, much more sophisticated method, which is consequently very time consuming and therefore expensive. Needs good as-built records.
c *Time slice windows analysis* is retrospective analysis using 'snapshots' taken throughout the project execution.
d *As-planned vs. as-built analysis* has the advantage of creating an accurate and traceable evaluation of the progress of the works.
e *Longest path analysis* is used to calculate the 'latest' earliest finish time of the project.
f *Collapsed as-built analysis* uses the as-built programme as a baseline and, by removing all of the delays from the programme, to ascertain when the project would have been completed 'but for' these delays. Also needs a critical path to be calculated in the as-built programme.

On large projects, it may be necessary to concentrate on key areas of the project where the works are affected. In complex situations, or where records have not been so comprehensive, it may be necessary to investigate the construction logic and explain the intended sequence of events and any departure from that sequence. This forensic analysis of the key areas of delay will usually identify the actions taken, progress made, reasons for the delays encountered, together with the actual cause of the delay.

In the first instance, it will be necessary to analyse the facts of the delay and establish that the delaying event in question was not something that was included within the contractor's risks. There must be a baseline programme and contemporary records to show that the event occurred as there will be a need to show that the event actually affected progress to the works and also caused a delay to completion, since any delay to an operation which is not on the critical path will merely take up float on the project and therefore not delay the final completion.

The type of delaying event and the timing of the analysis will affect the outcome of the assessed delay to completion. For example, a prospective delays analysis would be to look forward to try and predict what the effect of an event would be on the overall project, whereas a retrospective analysis would be used only where the event and its effects have already happened and would obviously have more certainty in its assessment.

The form of contract should, but rarely does, specify the method to be used and the required timing of the analysis, which will depend to a large extent on the other provisions within the contract, the information available and the outcomes and proof required. The more detailed the information available, the more comprehensive and accurate the method of analysis will be.

13.5.1 Impacted as-planned

This method requires the identification of the baseline programme and the modelling of the event in question, with the use of planning software. The modelled event is then added into the programme and the completion date is re-calculated. Usually, but depending upon the logic in the planned programme, the completion date will change following the reschedule and it is the movement in the completion date that represents the delay caused by the imported events using this technique.

Therefore, the revised end date and the original end date are compared and the difference reported. Again, this is a fairly quick and easy method which can be used where there is limited site information available and the critical path remains largely unaffected but for the event in question. However, this method ignores the progression of the works, the changes in sequence and timing or the effects of concurrency or changed logic. The method can be used either prospectively or retrospectively but its outcomes are not considered as reliable as with other methods.

13.5.2 Time impact analysis (TIA)

This method requires the individual study of each claimed event causing delay and the original programme to be updated for each delaying event. Each update requires the construction logic to be examined for every change with durations re-evaluated and new activities created. The updated programme is re-modelled using each event and therefore may result in both a new critical path and an extended period of time. This iterative process is repeated for each updated event in chronological order and provides a record of events as they unfold by putting the sequence of works into context to the project as a whole. This is a highly sophisticated method of analysis as it identifies both the cause and effect whilst taking account of progress, resources and logic.

TIA is considered as the most accurate method because it uses all planned, progress and as-built information whilst considering any inadequate progress or concurrent delay. It also takes any changes in methodology or re-sequencing and acceleration into account and can cope with multiple key dates or milestones. However, the disadvantages are that it can be an extremely slow and expensive method of analysis which requires high-quality information and produces a complicated and cumbersome reflection of the events on site. The analysis itself can produce a large volume of material which may be difficult to communicate to lay clients and therefore requires a great deal of explanation and presentation. However, this is an excellent technique for assessing interim extensions of time, negating the need to 'wait and see' what happens.

There are eight major steps in this analysis, to understand the scope change and its impact on a completion date:

1 Develop and approve a sequence of new activities as a sub-network (termed a 'fragnet') to model the extra work.
2 Obtain the most recently updated approved schedule. An older schedule update should not be used as it may not reflect the current work status.
3 Set the durations of the extra work activities (fragnet) to zero.
4 Insert the approved durations into your fragnet (schedule of new work) to re-calculate the critical path.
5 Identify the activities used to measure the delay. If the new critical path does not include the new fragnet, then clearly it will not delay the project, although there may still be a resource availability issue, which can be checked separately.

6 Determine the correct time impact of delay.
7 Determine the actual dates of the delay.
8 Eliminate any delay that was already in the schedule (the difference between the completion date with the fragnet and the recently updated CPM schedule will be the delay impact due to the extra work).

13.5.3 Time slice windows analysis

This is also a 'retrospective' analysis method and requires both a networked baseline schedule and good as-built data in order to allow for progressive updating. In simple terms, the baseline schedule is updated at regular intervals throughout the project – usually monthly – and each update will provide a 'snapshot' of the project status at that point in time, from which two important pieces of evidence can be obtained. Firstly, the sequence of activities in the critical path to completion at each point of update, and secondly, the extent of any possible delay to completion established at that date.

This method can be very effective and reliable, provided that the snapshot schedules are an accurate reflection of the status of the works on the respective dates. There are three issues that can influence the reliability of the analysis:

- The baseline schedule must be both detailed and networked, so that the logic relationships within the schedule must facilitate reasonable and practical conclusions.
- The historic element of each snapshot schedule must be developed from accurate and detailed as-built data. If regular detailed progress information is not available then this method of analysis will not give reliable results, particularly regarding the route of the critical path.
- The future element of each snapshot schedule should be an accurate representation of the status of the works on the analysis date. The sequence of future work may or may not be similar to the sequence and logic of the original schedule.

13.5.4 As-planned vs. as-built

A quick, relatively inexpensive but also approximate method, which simply compares the original planned programme with the as-built programme and equates the difference as being the delay encountered. This is a quick, cheap and cheerful method of delay analysis but can be a useful way to review delays and quickly establish the possible merits of a claim. They can be acceptable where the existence and effect of an event is indisputable and also impacts the critical path. The limitations of this method are numerous, in that it is only a retrospective analysis and can be easily manipulated to suit the preferred case. It cannot deal with concurrent causes of delay, acceleration or re-sequencing as it does not calculate the actual effect a delay has had on the works by making the assumption that the fault lies with the other party.

The as-planned vs. as-built schedule delay analysis is therefore a retrospective method of analysis which involves comparing the baseline, or as-planned, construction schedule against the as-built schedule or a schedule that reflects actual progress. This analysis method is typically utilised when reliable baseline and as-built schedule information exists, but updated schedules either do not exist or are not reliable enough to support an accurate delay analysis.

The actual analysis can vary from a simple graphical comparison to a more sophisticated procedure which considers the start and finish dates and relative sequences of the various schedule activities to a more complex analysis which compares the start and finish dates, durations, and relative sequences of the activities and seeks to determine the fundamental causes of each variance. The complexity of analysis generally depends on the nature and complexity of both the project and the issues being evaluated. Due to its relative simplicity, this method is preferred by subcontractors and also for global claims where there are multiple causes and effects.

13.5.5 Longest path analysis

This technique calculates the longest continuous path of activities through a project which will be necessary for completion as it is technically possible for the critical path to not actually be the longest path and the longest path activities to not actually show calculated critical float. The longest path is determined by the string of activities and logic that push the project to its latest early-finish date. It is calculated by first determining the latest possible early-finish date and establishing the activities which constitute the project's longest path. For complete accuracy, longest path analysis should take place absent of constraints, resource leveling, and / or interruptible activities.

13.5.6 Collapsed as-built

This is a more accurate and detailed method of analysis, also known as 'as-built but for'. This method is carried out when the works are completed by using the as-built programme as a baseline and by removing all of the delays from the programme to ascertain when the project would have been completed 'but for' these delays. The analysis can be crude, by removing all delays at once, or can be made more sophisticated by removing each single delay in reverse chronological order in a stepped process. The advantages of this method are that it is factually based and easily understood, it can be used where there is no effective planned programme and it can also be used to demonstrate both excusable and non-compensable delay (e.g. concurrent delay) whilst demonstrating the cause and effect of an event and its sequence in the overall as-built programme. However, the analysis is complicated; it requires accurate and complete as-built data with the reconstruction of logic which may be challenged as subjective and lacking in the acknowledgement of actual re-sequencing or accelerative measures.

13.6 Summary and tutorial questions

13.6.1 Summary

We have seen that, as with variations (see Chapter 6), all construction projects are generally subject to some form of delay and / or disruption to the original planned schedule. Well-managed projects are brought back on track so that the cause of delay and disruption does not affect the overall completion, which clearly needs teamwork between the employer, PM / engineer and contractor working together to find solutions. Poorly managed projects and those where there is a significant blame culture will not be so fortunate, so it is absolutely vital that good records are kept throughout

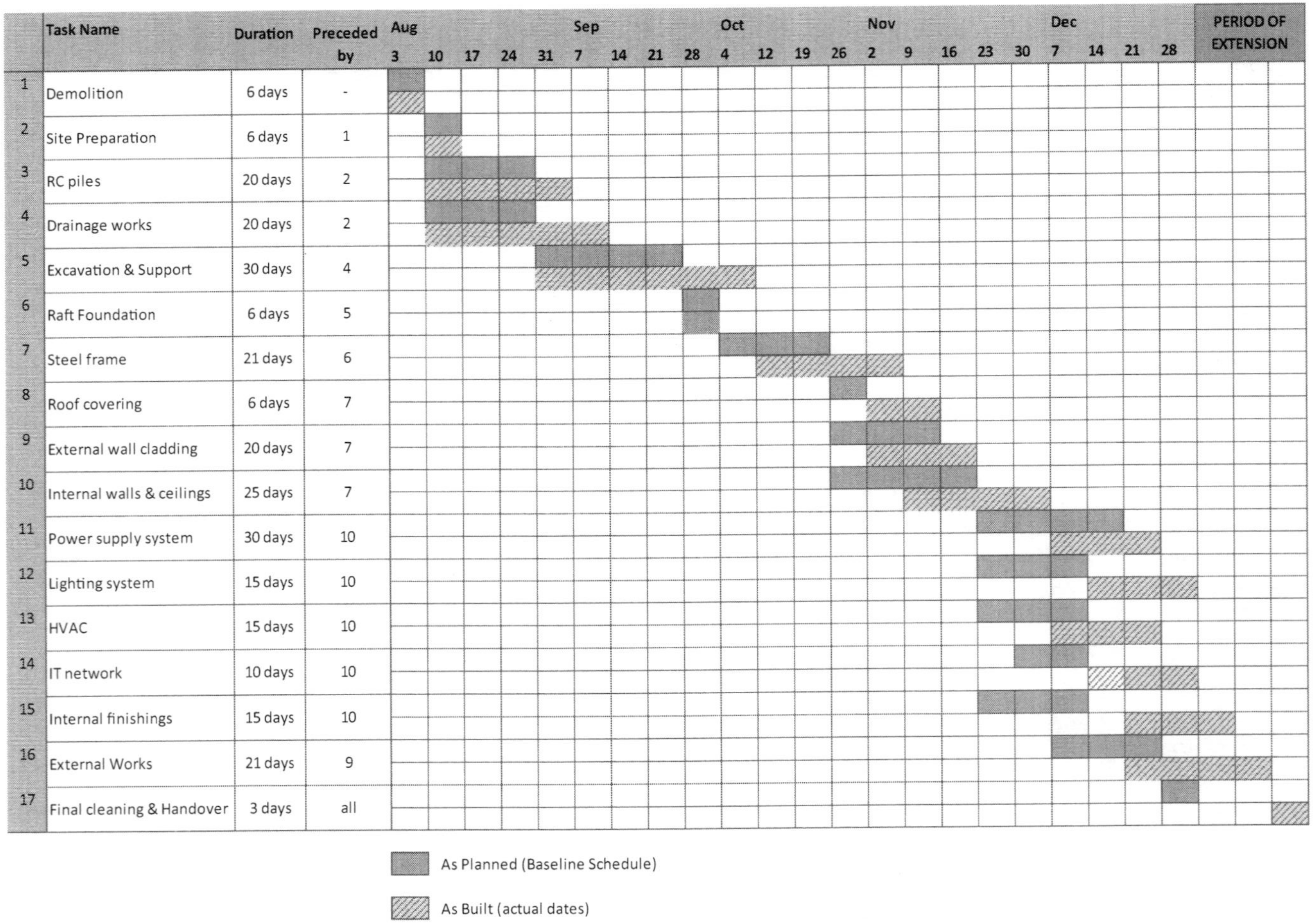
Task Name
Duration
Preceded by
Aug
Sep
Oct
Nov
Dec
3 10 17 24 31 7 14 21 28 4 12 19 26 2 9 16 23 30 7 14 21 28
PERIOD OF EXTENSION
1 Demolition 6 days -
2 Site Preparation 6 days 1
3 RC piles 20 days 2
4 Drainage works 20 days 2
5 Excavation & Support 30 days 4
6 Raft Foundation 6 days 5
7 Steel frame 21 days 6
8 Roof covering 6 days 7
9 External wall cladding 20 days 7
10 Internal walls & ceilings 25 days 7
11 Power supply system 30 days 10
12 Lighting system 15 days 10
13 HVAC 15 days 10
14 IT network 10 days 10
15 Internal finishings 15 days 10
16 External Works 21 days 9
17 Final cleaning & Handover 3 days all
As Planned (Baseline Schedule)
As Built (actual dates)

Figure 13.2 As-planned vs. as-built analysis.

the construction period of what happened and when. The early chapters of this book outline some of the techniques to be used for keeping these records.

Delay and disruption are not the same, and whereas delay can be defined in a relatively straightforward way, disruption is far more complex and therefore requires more analysis and records to prove any entitlement. Some of these records relate to the productivity and efficiency of the resources used, which can be extremely sensitive and confidential to a particular contractor. Therefore, most contractors will always try to show an entitlement to delay rather than disruption.

The contractor's primary obligation is to complete the scope of works and they have undertaken to complete the original scope of works in the original time period. Employers will invariably vary this scope, which may require additional time, unforeseen events occur and before you know it, the original completion date is unachievable.

The contractor has an obligation to mitigate any delays and some may have been their fault anyway (culpable delays). Where there is a valid case for claiming additional time and costs, the contractor will be required to forward a Notice of Claim and substantiate any claim for entitlement with valid and accurate contemporary records. Many of the delay events will overlap each other and the contractor will not be entitled to time or costs for culpable delays, but what happens with concurrent delays? This has been the subject of considerable legal opinion and in many cases the contractor loses out even though the employer was also at fault during the period of concurrency.

For this reason, many contractors submit what are termed 'global claims' – where several events are bundled together and submitted as one aggregate claim. This again is a major area of claim analysis and the reader is encouraged to refer to more specialist texts for further details.

In terms of delay analysis and evaluation, which analysis technique to use depends on the availability of an as-planned programme, whether it is a network, schedule or a mere list, as well on the quality of the records that have been kept during the progress of the project (i.e. contemporary records or contemporaneous records – same thing but depends on how clever you want to look).

Delay analysis is not an exact science and some, if not many, assumptions are always required. Some of the delay analysis techniques are prospective (i.e. looking forwards – before the event) and others are retrospective (i.e. looking backwards – after the event). The prospective techniques (impacted as planned and time-impact analysis) are best used during the project as they predict the likely impact of delays on the remaining unfinished work. The main difficulty with the prospective techniques is that they actually calculate the effect of delays on the programme rather than on the project. So, if the programme is wrong or not sufficiently detailed, or includes poor logic links, then the conclusions generated will be not be robust and may also be unreliable (garbage in – garbage out). The impacted as-planned technique is entirely theoretical, as is time impact analysis, but the theoretical element diminishes with time because it has the advantage of taking account of progress of the work.

It is critical to understand the logic behind each approach together with the respective strengths and weaknesses before deciding which technique to use in practice. Whilst the techniques can be applied to give a precise conclusion or output, it is unlikely that a precise answer will ever be absolutely correct due to the assumptions

Table 13.1 Summary of delay analysis techniques (Source – Rider 1 to the SCL Delay and Disruption Protocol: July 2015)

Method of Analysis	*Analysis Type*	*Critical Path determined*	*Delay Impact determined*	*Requires:*
Impacted As-Planned Analysis	Cause & Effect	Prospectively	Prospectively	Logic linked baseline programme. A selection of delay events to be modelled.
Time Impact Analysis	Cause & Effect	Contemporaneously	Prospectively	Logic linked baseline programme. Update programmes or progress information with which to update the baseline programme. A selection of delay events **to be modelled.**
Time Slice Windows Analysis	Effect & Cause	Contemporaneously	Retrospectively	Logic linked baseline **programme.** Update programmes **or progress information** with which to update **the baseline** programme.
As-Planned versus As-Built Windows Analysis	Effect & Cause	**Contemporaneously**	Retrospectively	Baseline programme. As-built data.
Longest Path Analysis	Effect & Cause	Retrospectively	Retrospectively	Baseline Programme. As-built programme.
Collapsed As-Built Analysis	Cause & Effect	Retrospectively	Retrospectively	Logic **linked** as-built programme. A selection of delay events to be modelled.

required and the flaws inherent with all of the methods. This range of probability should always be taken into account when evaluating the answers.

There are several important considerations for choosing the most appropriate analysis technique, as each contractual claim is unique and deals with different contract requirements, situations, level of documentation, complexities, legal jurisdictions and dispute resolution forums, among other things. The selection of a particular analysis method should be based on professional judgement and diligent factual research and evaluation.

Whether the delay analysis is carried out during the course of the project or after completion, it is always advisable to use the same software package that was used to develop the original construction programme and/or updated programmes. Changing software packages, especially at the end of the project, may not reflect the intent of the original network analysis or schedule.

13.6.2 Tutorial questions

1. Why do construction contracts need express provisions to extend the date for completion?
2. What is the situation if there was no extension of time provision in the contract?
3. What is the difference between 'delay' and 'disruption'?
4. Give appropriate examples of a 'condition precedent'?
5. Comment on the difference between acceleration and recovery.
6. Why is it essential for the contractor to keep accurate contemporary records during construction?
7. Discuss the major difference between prospective and retrospective delay analysis techniques.
8. What are the major advantages and disadvantages of:

   a. As-planned vs. as-built?
   b. Time impact analysis?
   c. Impacted as-planned?
   d. Collapsed as-built?

14 Adjudication, dispute boards and ADR

14.1 Adjudication

14.1.1 What is adjudication and why do it?

Adjudication is now a dispute resolution procedure which has been fully consolidated in the UK construction industry since it was first introduced by the Housing Grants Construction and Regeneration Act 1996 as a direct consequence of recommendations in the Latham Report – *Constructing the Team* – in 1994. The Act made adjudication a mandatory 'first step' in all construction disputes and the adjudicator's decision, although 'quick and dirty', was invariably upheld when challenged in the courts; however, the parties were free to open up the decision again in arbitration proceedings should they feel the need to do so. It was designed as a procedure for resolving disputes without resorting to lengthy and expensive court procedure.

Originally the intention of the Construction Act was that the process would be fairly informal without the need for lawyers, judges etc; but it has now developed much more formality with parties serving detailed submissions, Witness Statements and Expert Reports.

In terms of costs, the Construction Act makes no mention of how costs should be divided between the parties, and many employers included a clause in their contracts to the effect that the contractor would be responsible for all costs (for both parties) of any adjudication procedures irrespective of whether they were successful or not. Because of this, changes to the Act in 2011 provided that any contractual provision which attempts to allocate the costs of adjudication between the parties will be invalid unless it is made after the adjudicator is appointed. This applies to agreements for the allocation of the adjudicator's fees and expenses as well as any agreements regarding who is to pay the parties' own costs.

Regarding the fees and expenses of the adjudicator, the parties are jointly and severally liable to pay the adjudicator a reasonable amount in respect of fees for work reasonably undertaken and expenses reasonably incurred. This means that both parties can be sued for these fees, or that either party may be pursued for the whole amount. The adjudicator will invariably have a set hourly / daily / weekly rate but, if there is any dispute, an application can often be made to the court for determination.

14.1.2 Advantages and disadvantages of adjudication

The following table is based on UK law.

Table 14.1 Advantages and disadvantages of construction adjudication

Adjudication in Construction	
Advantages	*Disadvantages*
1 Statutory right (in UK). Adjudication will apply even if the contract does not provide for it – The 'Scheme' will apply.	1 Rough justice – Given the time constraints adjudication can sometimes be seen as quick and dirty as the responding party may only have a matter of 2–3 weeks to prepare a defence to the dispute.
2 Confidential – As the proceedings are conducted in private the dispute can be resolved without being heard in open court.	2 Rough justice 2 – incorrect decisions may still have to be honoured until opened up again in arbitration.
3 Costs – Decisions in adjudication will, in the majority of cases, be a fraction of the cost of pursing a dispute through arbitration or the courts.	3 Legal costs – Unlike the courts, the adjudicator may not have the power to order the losing party to pay the winner's legal costs.
4 Speed – An impartial decision can normally be obtained within weeks whereas arbitration or litigation may take months to reach a decision.	4 Possibility of many adjudications in a large project, all taking up valuable project time and leading to more complex arbitrations if those proceedings are used.
5 Flexibility – The parties may, on agreement, extend the time limits for response depending on the complexity or volume of material to be considered.	
6 Written reasoning – The adjudicator may provide written reasoning for the decision.	
7 Final and binding – The decision of an adjudicator is normally binding unless appealed to arbitration or litigation.	
8 Payments can be enforced quickly, thus assisting the contractor's cash flow (if they win).	

14.1.3 The adjudication process

The adjudication process begins when the party referring a dispute gives written notice of its intention to do so. This Notice of Adjudication should briefly set out the following:

- a description of the nature of the dispute and the parties involved;
- details of where and when the dispute arose;
- the nature of the remedy being sought;
- names and addresses of the parties to the contract, including addresses where documents may be served.

The Notice of Adjudication is therefore the first formal step in the adjudication procedure. Except for the minimum information detailed above, there is no particular requirement regarding the form of the document – remember that adjudication is supposed to be an informal and quick process.

Appointment of the adjudicator

Following service of the Notice of Adjudication, the next step is to actually appoint the adjudicator. The appointment of an adjudicator must be secured within seven days from service of the Notice of Adjudication and the parties can agree on an individual to act as the adjudicator or, if agreement cannot be reached, the party who referred the dispute to adjudication may make an application to an Adjudicator Nominating Body (ANB). This is usually done by completing a simple form and paying the obligatory fee. On receipt of a request to nominate an adjudicator, the ANB should communicate their selection to the party who referred the dispute to adjudication within five days of the request. In the event that an ANB fails to do this the whole process must begin again.

The Referral Notice

The Referral Notice must be served within seven days of service of the Notice of Adjudication. This document sets out in detail the case of the party who is referring the dispute to adjudication and it should be accompanied by all documentation necessary to support the claim together with Expert Reports (if any) and Witness Statements. It is important to ensure that the referring party is in a position to serve this notice, so it should ideally be prepared before the Referral Notice is given – there have been instances where the ANB has appointed an adjudicator only 24 hours before the seven-day period expires, in which case the adjudicator will need the notice within a day. A copy should be sent to the other party at the same time.

Timetable involved

The Construction Act sets out a tight timetable of 28 days after the service of the Referral Notice for submission of a response and for the adjudicator's ultimate decision. However this may be extended with the consent of the adjudicator. The rationale behind the process was to obtain quick and cost effective results which are of a binding nature unless reviewed by litigation or arbitration. This clearly relies on timescales being tight. Tight timescales also mean that the risk of an 'ambush' is lessened, where one party serves enormous volumes of documents, which clearly cannot be evaluated fully in the time available.

Responding party's response

This is essentially the other party's defence, and is required to be served within seven days of the Referral Notice. Requests for this to be extended to 14 days are usually agreed. The Construction Act does not demand a response or further submissions – the need for these is purely a matter for the adjudicator.

The decision

The adjudicator is required to reach a decision within 28 days of service of the Referral Notice. This period can be extended by a further 14 days if the party who referred the dispute in the first place agrees, or can be further extended if both parties agree.

The decision is final and binding, providing it is not challenged by subsequent arbitration or litigation. As mentioned in table 14.1 above, the parties are obliged to comply with the decision of the adjudicator, even if they intend to pursue court or arbitration proceedings. In the majority of adjudicators' decisions the parties accept the decision; however, if they choose to pursue subsequent proceedings the dispute will be heard afresh – not as an 'appeal' of the adjudicator's findings. Once a particular issue or dispute has been subject to an adjudication, it cannot be included in further adjudication proceedings.

14.1.4 International aspects of adjudication

Countries having similar legal systems to the UK (i.e. based on common law) have tended to replicate the above procedures – Australia, New Zealand, Singapore and Malaysia all now have legislation which provides for adjudication of construction disputes in the first instance. The mechanics of the legislation naturally varies depending on the detail of the legal system but they all share the requirement to enable a quick, binding dispute resolution process in order to avoid costly delays (most legal systems are notoriously slow and cumbersome), together with non-value added lawyers' fees.

The situation in the Middle East is different. The enormous inflow of petrodollars has led to increasing levels of development in recent years with a considerable amount of construction work, both in building and infrastructure projects as well as oil and gas process engineering plants. All of the Gulf Cooperation Council (GCC) countries – Kuwait, Saudi Arabia, Bahrain, Qatar, UAE and Oman – have benefitted from this development, although Dubai is probably best known due to the 'statement' projects such as the Burj Khalifa and the Palm man-made islands. Despite a nosedive in construction activity following the credit crunch recession in 2008–2009, the region has continued to grow and design has started on a major high speed rail connection from Kuwait in the north to Oman at the south end of the Gulf, linking all countries and major cities en route.

Due to the cultural processes and the relative speed of development of these countries, the local legal system has been unfamiliar with complex construction contracts and local clients have not always been keen to use international (i.e. foreign) arbitration tribunals to instruct them what to do. Therefore, the countries with the highest incidence of contractors (i.e. Dubai and Qatar) have set up their own arbitration centres, so that decisions are at least made in-country and in accordance with their own practices and procedures. Unfortunately, adjudication has not been accepted in the same way in the Middle East, which is perhaps due to the historical lack of this type of quick and dirty procedure both in the local market and also in the countries where many of the international contractors are based. On the other hand, dispute boards have been used in the region for some time, primarily because they are included in the FIDIC Standard Form of Contract 1999 edition, at Clauses 20.2 to 20.4, as Dispute Adjudication Board.

14.2 Dispute boards

14.2.1 What is a dispute board?

The term dispute board is used to describe a dispute resolution procedure, or more accurately a dispute avoidance procedure, which is normally established at the beginning of a project and remains in place throughout the life of the project – or at least the life of the contract under which it was set up. It normally comprises up to three members who are required to be familiar with the project as well as the contract documentation in order to provide informal assistance and recommendations so that disputes and disagreements can be settled before the issues get too entrenched in the parties' minds. The dispute board may also provide binding decisions on disputes if that is written in to the agreement. The members of the dispute board are often paid a monthly retainer to be available as and when required, which clearly can cause difficulties for busy professionals as they cannot always drop everything else and respond to a request to attend the project, especially if it is in another country thousands of miles away.

The concept of dispute boards is not new but has been slightly revived in recent years due to the increasing use of adjudication across many parts of the world and also the use of dispute review boards (DRBs) which were originally developed for major projects in the USA. Also, several of the new contract variants (e.g. the FIDIC Pink Book, a version of the Red Book for multilateral development banks) include dispute boards as essential components of the project structure.

Dispute review board or dispute adjudication board?

Dispute adjudication boards (DABs) are provided for under the FIDIC suite of contracts as well as under the ICE Procedure and it is also possible to provide for a DAB under the ICC rules. The DAB issues decisions on disputes which will be binding on the parties on an interim basis, in much the same way as the adjudicator's decision described in section 14.1 above. Dispute review boards, on the other hand, do not issue decisions, but rather recommendations, which the parties are free to accept or ignore as they see fit since they have no binding effect on either party. There are, however, sometimes provisions in the contracts which may make a recommendation binding if certain steps are not taken by one or other of the parties. As ever, read the specific contract requirements.

The fact that the DRB's recommendations are generally non-binding has not had a major effect on the way they have operated in the USA or made them any less effective. As with an adjudicator, the members of a DRB will generally be technically savvy construction professionals rather than lawyers, so the recommendations will be of a more practical nature rather than purely legal or contractual.

The way in which a DRB approaches its task is also different from a DAB mainly because of the differences in their dispute resolution roles and make-up. A DRB, whilst generally taking into account the provisions of the contract and the applicable law, will also consider the interests of the parties and the project requirements. In contrast, a DAB will normally follow the strict provisions of the contract together with the applicable law and is likely to adopt a more formal procedure in case of any legal challenge. It is unlikely that a DAB would ever take into account the wider interests of the parties or the project and therefore its decision will be normally be purely based on the merits of the issue under consideration.

Or a combination of both? The ICC's combined dispute board

Rather than having to choose between a DRB and a DAB, the standard form of contract produced by the International Chamber of Commerce (ICC) also includes an option for a combined dispute board, or CDB. Under normal circumstances, the CDB will offer recommendations related to any dispute that is referred to it, but may also give decisions on disputes if it is asked to do so by the parties, or sometimes if they feel like doing it themselves anyway. This form of contract is not strictly a construction contract as it can be used across different industries; it should also not be confused with the Infrastructure Conditions of Contract (ICC) which is a UK-based suite of contracts for civil engineering works and was designed to supersede the ICE conditions, following the decision of the ICE (Institution of Civil Engineers) to support the NEC suite of contracts. That's hopefully cleared that up, then.

14.2.2 Advantages and disadvantages of dispute boards

Table 14.2 Advantages and disadvantages of construction dispute boards

Advantages	Disadvantages
Dispute Review Boards	
1 As the DRB only makes recommendations to the parties, the process is less adversarial. 2 Hearings are shorter and simpler, therefore less expensive. 3 Recommendations may address underlying issues rather than just the symptoms. 4 DRB members are normally senior, independent practitioners, therefore recommendations are seriously listened to by the parties. 5 Recommendations are disclosable in arbitration and / or litigation. If they are ignored, an arbitrator may want to know why.	1 Contractual procedures may allow the losing party to delay enforcement. 2 As they are recommendations, no enforcement procedures.
Dispute Adjudication Boards	
1 Decision can be immediately enforced even though only temporarily binding. 2 Early settlement of disputes. 3 See also advantages of adjudication in table 14.1 above.	1 Possibility of becoming more adversarial than necessary. 2 Decisions taken by a committee. 3 See also disadvantages of adjudication in table 14.1 above.

14.2.3 The dispute board process – avoidance or resolution?

The processes that a dispute board would use will depend on whether it is a 'standing' board or an 'ad hoc' board. A standing board will be constituted at the beginning of the project with specific named members and will serve until the completion of the project. An ad hoc board on the other hand will be constituted for particular disputes and will not have the same level of long term commitment or wider project knowledge. There are clearly major advantages to a standing dispute board in that it avoids the unedifying argument between the parties about who should be appointed to consider a particular dispute, and the members of the board will have developed a good knowledge of the project before considering the details of the dispute. There is also the issue of costs: standing boards are notoriously expensive to set up and maintain, as the members will mainly be drawn from senior experienced professionals and will be required to visit the project at regular intervals to maintain an appreciation and understanding of the project and any issues which may develop into disagreements and disputes. In this way, and as mentioned at the beginning of this section, dispute boards should be seen more as a *dispute avoidance* mechanism rather than dispute resolution.

The main dispute avoidance features are that, at any time, the parties to the contract may jointly request the DAB to give an opinion or recommendation on any technical or contractual issues on the project. If it has done its job properly, the DAB should be aware of any upcoming problems and will have put these problems on the agenda of upcoming meetings during their regular site visits.

As mentioned above, the board members must visit the site and meet with the parties and the PM / engineer at regular intervals and especially at times of critical construction events. All formal hearings will be conducted at the site if possible and a reasoned decision given to both parties in writing. The DAB may also investigate project circumstances and will have direct access to banks and other project investors. The multilateral development bank (MDB) version of the FIDIC Red Book normally insists on the DAB being constituted before the contract is approved.

For these reasons, the acronym DAB is often referred to as a dispute *avoidance* board, rather than a dispute *adjudication* board. Additionally, although the title dispute *review* board has been used in the discussions above, the initialism DRB can also be used to mean dispute *resolution* board, both of which imply a subtle difference in the requirement. Time will tell which of these terms become the settled definitions.

14.3 Alternative dispute resolution

Alternative dispute resolution (ADR) techniques are usually defined as those which are non-binding on the parties, where the dispute resolver assists the parties in resolving their disagreements. They are normally entirely private processes and conducted without prejudice to the rights of either party in taking the dispute to a more formal or binding procedure. In the UK, the Civil Procedure Rules (CPR) state that the courts will want to see that the parties have made a serious attempt to resolve their differences through ADR before commencing litigation, in an attempt to reduce the pressure on the court system and also to reduce the costs to the parties.

In the three ADR systems mentioned below, the dispute resolver usually works within a tried and tested method to allow each party to air their grievances and make their points and then works to establish common ground for the parties themselves to come to a settlement. In terms of cost, the parties normally share the cost of the mediator / conciliator / expert and cover their own costs, which should be minimal, as any 'hearing' will normally take place in one of the parties' offices. These are considered excellent methods of dealing with disputes as they positively encourage the parties to talk to each other and reflect on the wasted costs of digging heels in.

14.3.1 Forms of ADR in common use in the construction industry

Mediation

Mediation involves the services of one or more 'mediators' whose function is to assist the parties in settling a disagreement / dispute by direct negotiations between or amongst themselves. The mediator(s) will act as impartial participants in the negotiations, guiding and consulting the parties to arrive at an acceptable settlement themselves. The mediator cannot impose a settlement and should only guide the parties toward their own settlement. In this way, the parties 'own' the final decision and it is invariably the case that a good settlement is where both parties are dissatisfied with the outcome – this means that there has been substantial compromise on both sides.

The method normally adopted is that, after a plenary introduction, each party will take up residence in separate meeting rooms with the mediator liaising between the two camps. The mediator is not required to have a detailed understanding of the facts of the case or the contract provisions but will try to help each party to see the weaknesses of their own case and the strength of the other party's. This will hopefully ensure that a compromise is reached which will be confirmed in a formal written agreement to settle their differences. Unless it is a particularly complex issue, a mediation meeting should be able to be completed in two to three working days.

Conciliation

Conciliation is a similar procedure to mediation, but the conciliator takes a more proactive role in the discussions between the parties and is also required to have a more detailed knowledge and understanding of the facts of the case as well as the particular contract conditions. The conciliator will have come to an opinion on the dispute and will try to persuade the parties of this view, which is clearly some way short of issuing a decision, as in adjudication. It is for the parties themselves to come to the agreement and therefore own the outcome, in a similar way to mediation described above. Because the conciliator is effectively 'selling' a solution to the dispute, the whole process should be completed in a shorter timescale than mediation – depending of course on the skills of the conciliator.

Expert determination

In this method, an acknowledged expert is appointed because they possess a recognised knowledge and understanding of the particular issues in dispute. The appointed expert will then agree a procedure with both parties and will be presented with the position statements plus all supporting documents. The expert may decide on a 'documents only' determination of the issue, or may consult with each party separately. The expert is effectively an investigator, with responsibility to establish the facts and to decide on the merits of the parties' cases. Even though the dispute has been determined by an independent and acknowledged expert, it may still not be a binding decision, unless the parties have prior agreed that it should be.

14.3.2 Advantages and disadvantages of ADR

Table 14.3 Advantages and disadvantages of ADR in construction

Alternative Dispute Resolution (ADR) in Construction	
Advantages	*Disadvantages*
1 Helps to prevent disagreement or dispute going to litigation or arbitration – nipped in the bud.	1 No prior disclosure of documents to the other party.
2 More informal process, therefore more flexibility of procedures.	2 Non-binding decisions, therefore dispute may be opened up again later leading to wasted costs.
3 Faster decision making.	3 Not controlled by legal system.
4 High success rate – vast majority of ADR cases result in settlements.	
5 Non-adversarial – procedure based on discussions and negotiations.	
6 Private and confidential – no dirty washing in public.	
7 Lower costs in preparation and execution.	
8 Resolution kept within the construction industry – no need for expensive lawyers.	

Overall, it is considered by the industry practitioners that the advantages far outweigh the disadvantages stated above.

14.3.3 Selecting the most appropriate ADR mechanism

The suitability of the different forms of ADR depends on several factors, such as the nature and value of the dispute, the relative attitude and financial resources of the parties, the desired outcome and how the parties wish to represent themselves.

For ADR to work effectively, all parties must be willing to submit their disagreement to one or other of the ADR procedures and be serious about wanting to come to a settlement as, clearly, if both parties are not willing, then enforcement of a settlement will be impossible.

Adjudication is (at least in the UK) a legal obligation for dispute resolution and litigation is, of course, the only option where a legal precedent is necessary or an injunction is required. Arbitration is available if there is an arbitration clause in the contract between the parties, but before these 'big hitters' come in, many disputes can be resolved amicably by ADR and any of the forms of ADR will be worth considering where the cost of judicial proceedings is likely to equal or exceed the amount of money in dispute.

Where parties wish to preserve an existing relationship, then mediation or conciliation will be helpful. A great advantage of mediation is that the mediator is not bound purely to consider the obvious disagreement between the parties but can bring in other unrelated matters, provided they help the parties towards settlement.

Where available, short arbitration schemes offered by trade associations, government departments, utility regulators and ombudsmen schemes can all provide a cheaper alternative for redress providing that their terms of reference are satisfied.

Early neutral evaluation might be applicable in cases where there is a dispute over a point of law, or where one party may have an unrealistic view of their chances of success. For technical disputes with considerable factual evidence, mediation or expert determination may be the most effective, especially where there is a requirement to keep sensitive commercial information private. However, where there is a significant imbalance of power between the parties, mediation may not be the best course of action.

14.4 Summary and tutorial questions

14.4.1 Summary

Since its first introduction in 1996, adjudication has been a success story in the UK construction industry, although there were some initial teething problems caused by a lack of understanding, unwillingness to compromise and not least some glaringly incorrect decisions by adjudicators. Notwithstanding these issues, the 'quick and dirty' nature of adjudication has been recognised as being able to nip disagreements in the bud and help prevent them growing into larger and more costly disputes. Although there is a set procedure to follow, the whole process should be treated as relatively informal so that parties can keep good relations for the remainder of the project duration.

Dispute boards are also designed to resolve disagreements before they can potentially get out of hand. Depending on the contract conditions, they are used as dispute avoidance procedures or dispute resolution procedures – it is therefore for the individual project to decide on what the initialism DAB / DRB actually stands for.

There are also other forms of alternative dispute resolution which can act as a much quicker and much cheaper way of resolving disagreements, and mediation, conciliation and expert determination should all be seriously considered depending on the nature of the disagreement as well as the circumstances of the project.

14.4.2 Tutorial questions

1. Outline the principles of adjudication including, in your own words, the advantages and disadvantages.
2. What skills and experience should an adjudicator ideally possess?
3. Outline the essential differences between a dispute review board and a dispute adjudication board.
4. What are the main differences between a 'standing' dispute board and an 'ad hoc' board?
5. Under what circumstances would mediation or conciliation be used on a construction project?
6. Why is there a need for dispute resolution if both parties are contractually obliged to act in 'good faith'?

15 Arbitration and litigation

15.1 What is arbitration?

Arbitration is a form of formal and binding dispute resolution which does not require recourse to the formal courts of law. It is a consensual process in the sense that it will only apply if the parties agree that it should and normally will require that an arbitration clause is included in the contract between the parties, which will take effect when a dispute escalates beyond the point of being capable of amicable settlement. If there is an arbitration clause in the contract, the arbitration procedures must be used before commencing any litigation or formal court proceedings.

Arbitration has its origins in England and grew out of the international and local courts that were set up as an alternative to the royal court system in the middle ages. These 'courts' were set up in response to the demand by businessmen and merchants for an alternative system for the resolution of mercantile or commercial disputes, because the formal court system was considered to be slow, not well suited to commercial disputes and invariably inaccessible to parties resident abroad. A predominant feature of the new system was that the strict formalities should be waived or set aside to allow for the decisions to be made as soon as possible.

The practice of arbitration was eventually given a statutory basis in England when Parliament passed the first Arbitration Act in 1698. This legislation has since been continually refined and updated, with the latest statute being the Arbitration Act 1996 (the 'Arbitration Act'), which is the principal statute covering arbitration under English law and sets out the principles and procedures to be followed in all formal arbitration tribunals.

Despite various attempts to define arbitration, there is no single accepted definition, which may reflect the multi-faceted nature of the process and the way it has evolved over time. A useful enough definition can be found at the online Free Legal Dictionary:

> The submission of a dispute to an unbiased third person designated by the parties to the contract, who agree in advance to comply with the award—a decision to be issued after a hearing at which both parties have an opportunity to be heard.

Interestingly, the Arbitration Act does not contain a definition of arbitration. However, it does set out clear statements of principle regarding what is expected from arbitration.

Section 1 of the Arbitration Act provides some general principles:

- The object of arbitration is to obtain the fair resolution of disputes by an impartial tribunal without unnecessary delay or expense.

- The parties should be free to agree how their disputes are resolved, subject only to such safe-guards as are necessary in the public interest.
- Court intervention should be restricted.

Given its origins, there is often an expectation that arbitration will be quicker and cheaper than litigation but this is not necessarily the case. In a court litigation, the parties only have to pay a modest fee for the services of the judge and the use of court facilities and most of the true cost is met from public funds. In arbitration, the arbitrator (or arbitrators) are paid a professional fee for their services and as they are usually senior members of their profession, their costs are often significant, particularly in complex construction cases which can last several weeks or even months.

The market for good arbitrators with the relevant experience in construction law, technology and management is small and the services of the best arbitrators are increasingly in demand. One reason for this is the growth of international arbitration as businesses increase their presence in new global markets resulting in a proliferation of cross border disputes. The law of England and Wales remains the favoured choice for many international contracts where the parties may come from different countries, which means arbitrators qualified in English law are busy and it may be difficult to find convenient dates for hearings. Also, because it is a consensual process, arbitrators are often reluctant to interfere in the management of the proceedings, whereas the courts see case management as a means to keep control of costs and avoid delays.

The arbitration community is aware of these issues and arbitral institutions are working hard in attempting to reduce time and cost in arbitration. In addition, because there is a limited right of appeal in arbitration cases, this can also significantly reduce the time and costs of the process.

15.1.1 Contractual basis

Apart from the statutory basis for arbitration which is provided for in the Arbitration Act mentioned above, the other basis of arbitration is in the contract between the parties. As a consensual process, all the contracting parties must agree to submit whatever dispute is in question to arbitration. The rights and obligations of the parties to arbitrate their dispute will therefore arise from the arbitration agreement they have concluded. Most standard forms of contract include separate and optional arbitration agreements. If there is no prior arbitration agreement between the parties, this form of dispute resolution is not available.

15.1.2 Advantages and disadvantages of arbitration

Whilst arbitration is closely related to litigation, as a legally binding dispute resolution procedure, there are several key differences:

Confidentiality

An arbitration is heard in private. The tribunal, the parties and their representatives are the only people allowed to participate in the proceedings unless the parties and the tribunal agree otherwise. The parties can explicitly agree that the arbitration will be confidential, but if there is no explicit agreement, a duty of confidentiality will be implied into the arbitration agreement (under English law).

Flexibility

Unlike the Civil Procedure Rules (CPR), which govern how court cases are to be conducted, the Arbitration Act does not lay down a rigid arbitration procedure. The form of each arbitration will be different and will vary according to the particular characteristics of the case and what is agreed by the parties.

Party autonomy

The parties can chose where the arbitration is to take place and the rules to govern the procedure of arbitration. The parties also have the ability to choose the arbitrators – unless an agreement has been made beforehand, e.g. 'the Arbitrator will be appointed by the RICS, Law Society etc.'.

Finality of award

The arbitrator's decision is normally final and binding on the parties. Permission to appeal (to a court of law) may only be obtained in special circumstances and the grounds for otherwise challenging an arbitral award are very restricted.

Enforceability of award

An arbitrator's award can usually be enforced in the same way and as simply as a court judgment. Internationally, arbitration awards are widely enforceable due to various international conventions such as the New York Convention on the Recognition and Enforcement of Arbitral Awards concluded in 1958 (the 'Convention'). The Convention provides a regime for the recognition and enforcement of arbitral awards within Contracting States. More than 145 Contracting States have ratified the Convention. The Convention provides only limited grounds to refuse enforcement. Enforcement of an arbitral award under the Convention is generally considered an easier route to enforcement than enforcing a foreign court judgment abroad

15.1.3 Comparisons with adjudication

While the Housing Grants, Construction and Regeneration Act 1996 as amended (the 'Construction Act') does not define adjudication the aims and duties laid down by both are similar. Both Acts aim to promote the concepts of impartiality on the part of the decision makers, as well as expedition and avoidance of unnecessary expense. The powers given to the adjudicator and the arbitrator are very similar and neither is liable for their decisions unless they act in bad faith.

However, there are some notable differences, which include:

- All arbitrations are governed by the Arbitration Act, whereas only adjudication of disputes under construction contracts are governed by the Construction Act.
- The arbitration process is underwritten by a written contract, whereas as of 1 October 2011 construction contracts need no longer be in writing in order to benefit from the statutory right to adjudicate.
- Parties to construction contracts are entitled to devise their own scheme for adjudication but it must comply with the principles laid down by s.108 of the Construction Act. Parties to an arbitration have a great deal more freedom to manoeuvre.

- Unlike with an adjudicator's decision, there is no statutory time limit on an arbitrator to issue an award.
- Adjudication is a holding process. An adjudicator's decision is binding pending litigation, arbitration or subsequent agreement. An arbitrator's award is final and binding subject to appeal in very limited circumstances.
- An adjudicator's decision cannot be enforced as a judgment, whereas an arbitrator's award may be enforced in the same manner as a judgment or order of the court to the same effect.

Arbitration used to be the favoured form of dispute resolution for many members of the construction industry. This is less so today for two reasons.

Firstly, the majority of disputes (particularly more straightforward disputes) are resolved in adjudication. Secondly, some of the historic advantages of arbitration over litigation have been lost. The Technology and Construction Court (TCC) has now tailored itself to better meet the needs of the construction industry. The judges are construction specialists and it has become apparent that resolving a dispute through litigation in the TCC can be both quicker and less costly than resorting to arbitration.

However, arbitration remains a common forum for resolving construction disputes, and all major standard forms of construction contract contain arbitration clauses or an option to select arbitration. In particular, arbitration remains the favoured choice for dispute resolution in international contracts where neither party wants to submit to the jurisdiction of the local court.

15.1.4 The arbitration process

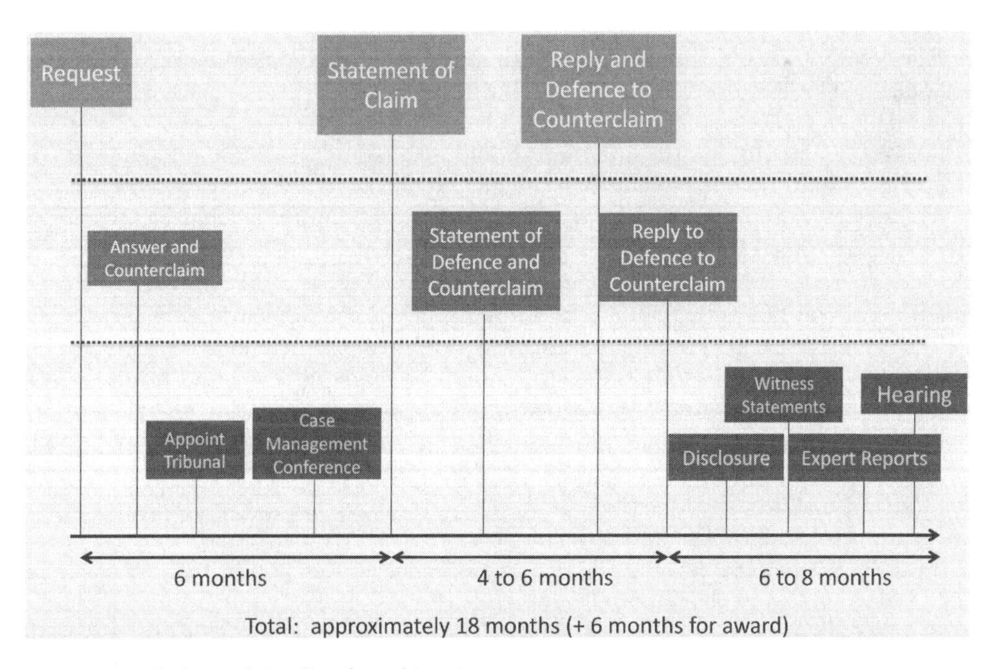

Figure 15.1 Estimated timeline for arbitration process.

15.1.5 International aspects

Arbitration is a truly international concept because it is a form of quasi-judicial commercial dispute resolution which sits outside the formal court structure in a particular country or legal jurisdiction. Clearly, the parties in an international project may come from different countries and after the project has been completed, may return to their country of origin. In this case, if they have been on the wrong end of an arbitration decision, the successful party will wish to get their money and will invariably consider legal action against the losing party. In these circumstances, the Convention on the Recognition and Enforcement of Foreign Arbitral Awards, also known as the New York Arbitration Convention or the New York Convention, is one of the key instruments in international arbitration. The New York Convention applies to the recognition and enforcement of foreign arbitral awards and the referral by a court to arbitration. There are over 150 countries who have signed up to this convention, which naturally gives some comfort to the successful party.

Another important document is the 1985 UNCITRAL Model Law on International Commercial Arbitration (amended in 2006), which is designed to assist countries in reforming and modernising their laws on arbitration in order to take into account the particular features and needs of international commercial arbitration.

15.2 What is litigation?

15.2.1 What is litigation and why use it?

Litigation is a formal legal process for enforcing a particular right or settling a dispute or controversy through the courts of the country holding jurisdiction over the matter(s) being considered. To litigate is to initiate proceedings between two or more opposing parties on matters which are ultimately settled: either by agreement between the parties or by a judge and/or jury in a court of law.

The term 'litigation' is not only used to describe proceedings in court, it can also include any number of activities before, during and after proceedings have been commenced and a lawsuit or claim has been issued.

The parties involved in the litigation should understand their case fully in terms of what the dispute is about, what each litigant expects to achieve and how to achieve it. The strengths and weaknesses, including the evaluation of the evidence on each of the matters claimed, should be considered in great detail in order to establish a case theory. There will then be developed an initial strategy regarding how the process should and will be conducted. There will be preliminary calculations in relation to cost and time for each stage before, during and after bringing the matter to actual trial. Each party, and latterly the court, should be made fully aware of those costs, which will (hopefully) assist in the various stages of the process including offers to settle or pre-action protocol. Nobody want to incur non-value added costs, but unfortunately, opinions can get hardened as disputes progress – therefore a clear, cold statement of cost may soften some of those attitudes and develop more of a willingness to compromise and settle.

In addition, the court has the power to consider the parties' conduct before proceedings and penalise any behaviour which the court finds unreasonable or uncooperative with adverse cost orders or other sanctions. The parties should be made aware that all conduct, both before and after proceedings are issued, including any offers to settle, will be considered and is likely to have consequences in terms of the allocation of costs. It is therefore important that the parties are reasonable in their behaviour in terms of the negotiations, the giving of information or the adherence to deadlines.

15.2.2 Pre-Action Protocol

Some matters of claim are covered by pre-action protocols which explain the conduct and set out the steps the court would normally expect parties to take before commencing litigation proceedings. As stated above, parties are usually encouraged to cooperate as far as practicable with each other by acting reasonably in exchanging information and documents, which may ultimately avoid the need to issue proceedings. Therefore, where there is no Pre-Action Protocol, it is appropriate to adopt one which is agreed by the parties.

In England and Wales, the Civil Procedure Rules now include a Pre-Action Protocol for Construction and Engineering Disputes. This sets the standards which the parties to a construction dispute or an engineering dispute are expected to observe before court proceedings are issued. The protocol encourages the parties to exchange information at an early stage and actively encourages using a form of alternative dispute resolution.

This protocol applies to all construction and engineering disputes, and covers consultants and professional negligence claims. A claimant may be able to avoid complying with the protocol if, by complying with it, the claim itself may be time-barred.

The typical stages of following the protocol include the claimant issuing a claim; the protocol specifies what information the Letter of Claim should contain. The contents of the claim form and any separate particulars of claim must be verified by a statement of truth and signed by a person with the appropriate seniority and knowledge. In the case of a company, this means a director, senior manager, CEO etc. It is therefore important for the claimant to check the claim very carefully, as any inaccurate or misleading statement may result in severe cost penalties or even contempt of court proceedings.

On receipt of the claim, the defendant must provide a full defence within 14 days, or 28 days if an acknowledgement of service has been issued within 14 days. This 28-day period of time can be extended by the consent of the parties up to a maximum of three months. If the defendant fails to acknowledge receipt of the Letter of Claim within 14 days the claimant is entitled to commence court proceedings without further compliance with the protocol.

Within 28 days from receipt of the Letter of Claim the defendant may raise an objection that either the court lacks jurisdiction, or that the matter should be referred to arbitration, or that the defendant named in the Letter of Claim is the wrong defendant. The objection should be in writing and should specify the parts of the claim to which the objection relates. It should also set out the grounds relied on and, where appropriate, it should identify the correct defendant, if the correct defendant is known. If the defendant fails to raise an objection at this stage, they are not precluded from raising an objection later. However, the court may take such failure into account when deciding the question of costs.

If the defendant raises an objection they are not required to send a Letter of Response in relation to the claim or those parts of it to which the objection relates. The defendant may withdraw any objection made at any stage; this must be in writing.

If the defendant withdraws an objection before proceedings are commenced, the parties are required to comply with the protocol. In such circumstances the Letter of Claim will be treated as having been received on the date on which the defendant withdraws the objection.

The defendant's Letter of Response should contain which facts and claims set out in the Letter of Claim are agreed and which are not agreed. Where the defendant does not agree with the facts or claims set out in the Letter of Claim, the objections should be clearly explained in the letter. If the defendant accepts a claim in whole or part, the Letter of Response should state whether the damages, sums or extensions of time claimed are accepted or rejected. If they are rejected the defendant should explain why they reject them. Where the defendant alleges contributory negligence against the claimant, the Letter of Response should contain a summary of the facts relied on. If the defendant intends to make a counterclaim then they should provide the same information in a Letter of Claim, naming any experts already instructed by the defendant, listing the evidence on which they intend to rely, and identifying the issues to which that expert's evidence will be directed.

The claimant then has a further 28 days from the date of receipt of the defendant's response to respond to any counterclaim. Again this period of time can be extended with the consent of the parties for up to a maximum of three months.

The parties must then have a 'without prejudice' meeting within 28 days of receipt of defendant's Letter of Response, or after the claimant's response to the counterclaim. This is so that the parties have an opportunity to reach an agreement on the main issues of the dispute, to identify the root cause of disagreement in respect of each issue, and consider whether the issues might be resolved without the need for court proceedings.

The parties are required to consider whether some form of alternative dispute resolution procedure would be more suitable than the commencement of proceedings. If the parties are unable to resolve the dispute they are expected to try to reach an agreement on matters relating to expert evidence, disclosure of documents and the conduct of proceedings in order to minimise delay.

The protocol expects the parties to act reasonably to keep costs proportionate to the complexity of the case and the amount of money involved. The protocol must not be used as a tactical device to secure an advantage for one of the parties or to generate unnecessary costs. If a party does not comply with the protocol, the court has discretion when it comes to awarding costs.

Once the process of pre-action is completed and if the parties are still unable to come to a settlement, the claimant should issue a claim. This is made usually by way of a claim form, clearly written and verified by a statement of truth, confirming the facts stated to be true. The claimant must check the claim very carefully, as any inaccurate or misleading statement could result in severe cost penalties or contempt of court proceedings being brought against the claimant.

A court fee will be payable in order to issue the claim, its value depending on the monetary value of the claim. Once issued by the court, the claim must be served on the defendant who then has 14 days to acknowledge the claim and 28 days to provide a full defence. From there on, either party may request further information or clarification from the other in relation to any matter in dispute. The speed and manner in which the claim is dealt with is then controlled by the court using its case management powers and any party not adhering to the timetable set will incur severe cost penalties.

15.2.3 Pre-trial proceedings and case management

Once the defence has been filed, the case will be allocated to a particular track by the court. This will be one of the following:

- the small claims track: for claims generally less than £10,000.00;
- the fast track: for claims less than £25,000.00 or for larger straightforward claims;
- the multi-track: for claims more than £25,000.00 or for more complex cases.

Each track contains different provisions regarding costs and timetable and the allocation by the court will be decided on the merits of each case. Usually, in construction cases the matter will be allocated to the multi-track in the Technology and Construction Court.

Directions

In the TCC, the claimant must make an application for directions for how the case will proceed within 14 days of the filing of the defence or acknowledgement of service, whichever is the earlier. In the directions hearing, the first case management conference will take place in consultation with the judge. If the claimant does not apply for directions, then any other party may be able to apply for the claim to be struck out or dismissed, otherwise the judge has the discretion to fix a directions hearing.

After notifying the parties of the time and date of the hearing, the court will send a first case management questionnaire and case management directions form for the parties to complete. Failure to complete the forms within the time stipulated will result in sanctions by the court. The parties are encouraged to agree directions in advance in order to further the overriding objective.

A date for the trial is usually fixed by the court at the directions and case management hearing; any preliminary issues will also be considered and prioritised as appropriate. The court will also give case management directions on how the case should proceed to trial with a date fixed for a pre-trial review.

Summary judgment

In the early stages of the case it will be important to consider and apply for summary judgment. The court may give summary judgment, without trial, if it is considered that the claimant has no real prospect of success or if the defendant has no real prospect of defending the claim and there is no other compelling reason why the trial should be held. In some cases, even where a party has a good case, if that party is not prepared, has not given sufficient particularisation in the pleadings or has not provided supporting evidence, it risks having summary judgment against it.

In such a situation, the parties should be able to give a detailed account of costs expended to date, and 24 hours before the hearing. If a summary judgment is unsuccessful, the court's case management directions will be issued, and the parties should be able to provide details and availability of witnesses, experts' evidence requirements and disclosure requirements. The parties should also provide an explanation of which other means of dispute resolution have been used or considered and if not, why not, together with any timetable suggestions.

Disclosure

Disclosure is a process whereby each party discloses to the other exactly what documents it has which are relevant to the case. It is the responsibility of both parties to disclose to the other any documents which would support or undermine their own or the other party's case. This is even the case where documents are confidential and parties must carry out a reasonable search for such documents. This duty, to disclose documents, continues until the proceedings are over. Parties may be able to object to producing a document, but only where that document is privileged or if it would be disproportionately costly to produce or if there is no significant reason to produce it.

Disclosed documents are sent within a list to the other party which must be signed by a representative of the party disclosing, certifying that they understand the responsibility to disclose documents and that this responsibility has been complied with. There must also be a statement as to the extent of the search carried out. Where a party wishes to inspect a document, written notice must be supplied with a corresponding right to do so being given within seven days of the date of the notice. The court may make an order for a specific document to be disclosed or inspected where a party has previously tried to exclude it.

Offer to settle

Within most litigation processes there is a culture in which disputes are managed and determined quickly with various mechanisms designed to encourage the parties to behave reasonably, exchange information and arrive at a settlement. Under the UK CPR there is an effective tool for promoting settlement. Part 36 offer or payment rules provide the opportunity for either a claimant or defendant to make an offer of settlement. Under the procedure, they offer a special status which cannot be sensibly ignored.

Where a defendant makes a Part 36 offer or payment (previously termed a 'Calderbank' offer) and this is accepted by the claimant within 21 days, the process will be stayed on the terms of the settlement offer and the costs of the claimant will be paid by the defendant. This rule exerts pressure on the claimant because, if at trial, they fail to obtain a greater or more advantageous outcome than the offer made, they will be compelled to pay the defendant's costs (as well as their own) after the expiry of the offer made. Therefore, the earlier a Part 36 offer is made, the greater the pressure placed on the claimant. In the opposite scenario, i.e. where the claimant makes an offer to settle, the cost consequences could be even more severe if the claimant achieves a more advantageous award as the defendant runs the risk of being ordered to pay interest up to a maximum of 10 per cent on top of the base rate. A defendant must therefore consider Part 36 offers very carefully.

Evidence generally

Any fact which needs to be proven by the evidence of witnesses is generally proven at trial by oral evidence, whereas in any other hearing it will proven by evidence in writing. In some cases, a court may deal without a hearing. Where a judge agrees that an application is suitable for consideration without the need for a hearing, it will give directions as to the evidence which should be filed.

Where one party intends to rely on a Witness Statement, the court will order for that statement to be served on the other parties by giving the order in which Witness Statements are to be served and whether these are to be filed with the court. A Witness Statement should be signed by the person making it and should only contain the evidence which that person would be allowed to give orally. It must be in the witness's own words and verified with a statement of truth: if there is no honest belief or if there is a false statement the witness may be prosecuted for contempt of court. All Witness Statements should be prepared as efficiently and accurately as possible, to avoid the costs associated with re-drafts as this will not normally be recoverable.

Each party should carefully consider the other party's Witness Statements and develop a line of questioning for cross-examination in line with the case theory.

Expert evidence

In addition to statements about the facts of the dispute, the construction industry invariably relies on evidence from acknowledged experts, who are appointed with the court's permission. Experts have a duty to the court over and above the party who instructs them and the court expects the expert to be impartial at all times. It is not unusual for each party to appoint their own expert and they often give differing expert opinions of the issue in question, so it is clearly for the court to decide which expert opinion is to be relied on, if any. With this in mind and where appropriate, the court will expect the parties to appoint a single joint expert and where the parties cannot agree, the court does have the power to direct that appointment.

The expert's evidence is usually given in the form of a written report, which must contain a statement with the substance of the instructions received, and a statement that the expert understands his duty to the court and that they have complied with that duty. Each party is permitted to put questions to the opposing experts, these questions should be for the purpose of clarifying the report and the answers will be considered as forming part of the report. The court may direct each party's expert to consult with one another and discuss the technical issues and, where possible, find mutual agreement. The experts are then expected to produce a report stating exactly which issues they agree on, which ones they do not agree on and a summary explaining the reasoning for any lack of agreement.

15.2.4 The hearing

Interim applications

A variety of interim applications may be made and the court has the power at any stage to exercise its case management powers and to give directions on how the conduct of the matter should proceed. Before the matter goes to trial, the Civil Procedure Rules provide that the parties must consider at each step whether the costs associated are proportionate to the matter in hand. If the costs cannot be justified, and where a party has not made or considered the appropriate interim application, they will probably be disallowed on any later assessment.

Where a party wishes to obtain an extension to a period of time already set by the court, there must be good reason as to why that time is needed. The onus is on the party who does not comply with a time period to submit an application for relief against any possible sanctions which might be imposed by the court.

Another possibility is where a party applies for a statement of case to be amended. This can only be done with the consent of all other parties or the permission of the court. The amendment should be accompanied with a verifying statement of truth and costs will likely be awarded against the amending party, as a form of penalty for not getting it right in the first place.

Where an application for security of costs order is made, it can only be made by a party in the position of the defendant (i.e. the defendant himself or the claimant making a defending counterclaim). This application is made where a defendant is confident that it can successfully defend the claim and obtain an order for costs but believes that the claimant will not be able to pay those costs. If this application is successful the claimant will be required to pay the sum of money into court, and if the security is not given the claim will be struck out.

Another interim application is for the remedy of interim relief, which includes the ability to stop a party from doing a particular thing or for a declaration that a particular state of affairs exists at that present time. Such orders can be made for property or assets to be preserved or detained, for interim payments to be made or to preserve evidence or access to property.

Applications for interim relief can be made at any time and may be made 'without notice' to the other party, if the court believes that there are good reasons for not giving such good notice. One example here is a freezing injunction which freezes the assets of a party before they are able to transfer those assets outside of the jurisdiction (i.e. to another country).

Applications for interim remedy should always be supported by evidence, unless the court orders otherwise.

Pre-trial review

In advance of the pre-trial review, the court will issue a pre-trial review questionnaire and directions form which the parties should complete, exchange and return to the court. If either of the parties fails to return the form within the stipulated period the court may impose sanctions (including striking out a defence or claim).

Trial

The trial bundle should be filed at court between three and seven days before the start of the trial, with the contents of the trial bundle being agreed between the parties wherever possible. At the commencement of the trial, both parties will give opening speeches and their principal evidence with written Witness Statements will be presented. Each party's evidence is then subjected to cross-examination whereupon the parties can then be re-examined. When all witness evidence has been presented, the parties will deliver their closing speeches. Following a period of consideration, the judge will then give judgment together with an assessment of costs.

Judgments

When the judge makes a decision, a written judgment will be issued and served on each of the parties. A party must comply with the judgment within 14 days of the date of the judgment, unless otherwise specified by the court.

15.3 Summary and tutorial questions

15.3.1 Summary

Whereas Chapter 14 covered the more informal dispute resolution procedures, which are also designed to be quick and relatively cheap, this chapter discusses the more formal and legally binding procedures, which can be extremely expensive for the parties as well as time consuming and stressful.

Although arbitration developed due to the complex and inflexible procedures of the law courts at the time, it has also developed its own complexities and operating procedures over the years which can be just as strict and unfriendly. The major difference is that it is a confidential procedure, so no dirty washing in public, and the arbitrator will be an experienced practitioner in their own right. Both procedures are final and binding with awards enforceable in the courts of all countries signed up to the New York Convention.

The subject of both arbitration and litigation needs much weightier tomes than this book to do proper justice to the complex procedures, so this is necessarily just a light touch introduction. In terms of how they compare to the informal procedures in Chapter 14, it may be useful to borrow the Hersey-Blanchard continuum of leadership styles as shown in table 15.1, showing that the types of dispute resolution methods falls reasonably neatly within the boxes.

This model was originally designed to illustrate the different leadership styles of a manager and show which would be most effective under different circumstances. Professional knowledge-based people invariably prefer their managers to be on the right hand side of the model, i.e. involving staff in any decision making, thus creating a sense of ownership. It is interesting to extend this model so that when parties are involved in the resolution of the disputes themselves, they are more likely to have a sense of ownership and therefore accept the result with better grace. Of course, in litigation, parties are forced by law to accept the result.

15.3.2 Tutorial questions

1 Outline the essential differences between arbitration and litigation.
2 Discuss the circumstances which must exist before parties can resort to arbitration.
3 What is the New York Convention?
4 Why is there a need for a specific protocol before litigation can take place?
5 What is the meaning of 'without prejudice'?
6 Define 'summary judgment' and 'disclosure'.

Table 15.1 Summary of methods of construction dispute resolution

Telling	*Selling*	*Participating*	*Delegating*
Litigation Arbitration Adjudication Expert determination	Conciliation	Mediation	Amicable negotiation

Part E

Ethics, fair dealings and anti-trust

This book is about construction contract management and both the people and organisations who will be undertaking that role on behalf of the employer will normally be independent, specialist consultancy organisations. There are many circumstances when the interests of the consultancy organisation may differ from the interests of the employer or indeed the project. Therefore rules of ethics and codes of conduct are extremely important in performing this service. For this reason, the subject is given a complete section in this book.

The following extracts are taken from the RICS regulations on ethics and professional standards. All professional members are expected to be able to demonstrate compliance with the five standards described below, which are at the heart of what it means to be a professional.

Act with integrity

Be honest and straightforward in all that you do. This is one of our five professional and ethical standards. This standard includes, but is not limited to, the following behaviours or actions:

- being trustworthy in all that you do;
- being open and transparent in the way you work;
- respecting confidential information of your clients and potential clients;
- not taking advantage of a client, a colleague, a third party or anyone to whom you owe a duty of care;
- not allowing bias, conflict of interest or the undue influence of others to override your professional or business judgements and obligations;
- making clear to all interested parties where a conflict of interest, or even a potential conflict of interest, arises between you or your employer and your client;
- not offering or accepting gifts, hospitality or services, which might suggest an improper obligation;
- acting consistently in the public interest when it comes to making decisions or providing advice.

Some of the key questions that you could ask yourself include:

- What would an independent person think of my actions?
- Would I be happy to read about my actions in the press?

(continued)

(continued)

- How would my actions look to RICS?
- How would my actions look to my peers?
- Do people trust me? If not, why not?
- How often do I question what I do, not just in relation to meeting technical requirements but also in terms of acting professionally and ethically?
- Is this in the interest of my client, or my interest, or the interest of someone else?
- Would I like to be treated in this way if I were a client?
- Do I promote professional and ethical standards in all that I do?
- Do I say 'show me where it says I can't' or do I say 'is this ethical'?

Always provide a high standard of service

Always ensure your client, or others to whom you have a professional responsibility, receives the best possible advice, support or performance of the terms of engagement you have agreed to. This is one of our five professional and ethical standards. This standard includes, but is not limited to, the following behaviours or actions:

- Be clear about what service your client wants and the service you are providing.
- Act within your scope of competence. If it appears that services are required outside that scope then be prepared to do something about it, for example, make it known to your client, obtain expert input or consultation, or if it's the case that you are unable to meet the service requirements, explain that you are not best placed to act for the client.
- Be transparent about fees and any other costs or payments such as referral fees or commissions.
- Communicate with your client in a way that will allow them to make informed decisions.
- If you use the services of others then ensure that you pay for those services within the timescale agreed.
- Encourage your firm or the organisation you work for to put the fair treatment of clients at the centre of its business culture.

Some of the key questions that you could ask yourself include:

- Do I explain clearly what I promise to do and do I keep to that promise?
- Do I look at ways to improve the service I provide to my clients?
- How can I help my clients better understand the services that I am offering?
- Am I providing a professional service for a professional fee?
- Would the client still employ me if they knew more about me and the workload I have? If not, why not?
- Do I put undue pressure on myself and colleagues (especially junior colleagues) to do more than we actually can?

Act in a way that promotes trust in the profession

Act in a manner, both in your professional life and private life, to promote yourself, your firm or the organisation you work for in a professional and positive way. This is one of our five professional and ethical standards. This standard includes, but is not limited to, the following behaviours or actions:

- promoting what you and the profession stand for – the highest standards globally;
- understanding that being a professional is more than just about how you behave at work; it's also about how you behave in your private life;
- understanding how your actions affect others and the environment and, if appropriate, questioning or amending that behaviour;
- fulfilling your obligations; doing what you say you will;
- always trying to meet the spirit of your professional standards and not just the letter of the standards.

Some of the key questions that you could ask yourself include:

- Do my actions promote the profession in the best light possible?
- What is the best way for me to promote trust in myself, my firm and the profession?
- Do I explain and promote the benefits, the checks and balances that exist with the professional services that I provide?

Treat others with respect

Treat everyone with courtesy, politeness and respect and consider cultural sensitivities and business practices. This standard includes, but is not limited to, the following behaviours or actions:

- Always be courteous, polite and considerate to clients, potential clients and everyone else you come into contact with.
- Never discriminate against anyone for whatever reason. Always ensure that issues of race, gender, sexual orientation, age, size, religion, country of origin or disability have no place in the way you deal with other people or do business.
- As much as you are able, encourage the firm or organisation you work for to put the fair and respectful treatment of clients at the centre of its business culture.

Some of the key questions that you could ask yourself include:

- Would I allow my behaviour or the way I make my decisions to be publicly scrutinised? If not, why not? If so, what would the public think?
- Are my personal feelings, views, prejudices or preferences influencing my business decisions?
- How would I feel if somebody treated me this way?
- Do I treat each person as an individual?

Take responsibility

Be accountable for all your actions – don't blame others if things go wrong, and if you suspect something isn't right, be prepared to take action. This standard includes, but is not limited to the following behaviours or actions:

- Always act with skill, care and diligence.
- If someone makes a complaint about something that you have done, then respond in an appropriate and professional manner and aim to resolve the matter to the satisfaction of the complainant as far as you can.

(continued)

(continued)

- If you think something is not right, be prepared to question it and raise the matter as appropriate with your colleagues, within your firm or the organisation that you work for, with RICS or with any other appropriate body or organisation.

Some of the key questions that you could ask yourself include:

- Am I approachable?
- Does my firm or organisation have a clear complaints handling procedure?
- Do I learn from complaints?
- Do I take complaints seriously?
- Am I clear about what the process is within my firm or the organisation that I work for about raising concerns?
- Have I considered asking for advice from RICS?

16 Ethics, fair dealings and anti-trust

16.1 Introduction

> Our most important company asset is our reputation for ethical behaviour, honesty and fair dealing. Reputation is a very fragile asset that can easily be destroyed by the actions of failures of one or more of us.

This is the opening statement of the code of ethics from a major international project management company. Most people would accept that there is a right way to do things under a particular set of circumstances – clearly most people would regard killing someone as wrong, but if the circumstances are that you are a soldier in an army at war, then killing is regarded as necessary and possibly acceptable. Fortunately, in construction contract management, although heated arguments are a frequent occurrence, it rarely descends into all-out warfare, so the limit of acceptable behaviour will be considerably below the actual taking of life.

A code of ethics therefore serves to establish a common understanding of the standards of behaviour expected of all employees in an organisation, as well as their service providers, consultants and contractors. It additionally provides a useful framework to assist in deciding on the most appropriate course of action when staff are faced with an ethical issue or dilemma. Just as important, a published code of ethics provides a degree of protection if the employee chooses a course of action which may be economically disadvantageous to the company – see the case study at the end of this chapter.

Adhering to ethical standards goes to the very heart of what is meant by a 'professional service' and construction projects bring together many professional service providers – architects, engineers, surveyors, construction managers etc. all aspire to this title because they are providing their services on behalf of their clients and should therefore put their client's (and the project's) interests before their own. The professional service provider may have obtained the work by negotiation or, more likely, by competition; but whichever is the case, they would not even be considered if their *reputation* was of poor quality.

The sections of this final chapter look at some of the issues which can affect this reputation, including the obligations, rights and responsibilities associated with those entrusted with managing construction contracts. The fact that they are entrusted with managing the *commercial* aspect of construction contracts makes the issue of ethical conduct even more crucial.

16.2 Professional obligations

16.2.1 Collusion and anti-competitive practices

Collusion involves secret or illegal cooperation or conspiracy, especially in order to cheat or deceive others. It is therefore clear to see that collusion is obviously an unfair business practice, whether as an informal agreement to share work between a group of companies, or as a more formal arrangement, such as a cartel. From 2004 to 2008, there was a major investigation by the UK government's competition agencies into collusive tendering (or bid rigging to give it a less euphemistic title) in the construction industry. This occurs when businesses, that would otherwise be expected to compete, secretly agree to raise prices or lower the quality of goods or services for projects tendered through a bidding process. Bid rigging conspiracies can take many forms, such as cover pricing, bid suppression, bid rotation and market allocation, all of which act against the interests of the employer.

Cover pricing – in some circumstances, a contractor may not wish to win a particular project, for example if all their resources are fully occupied for the foreseeable future. However, they may not wish to withdraw from a particular employer's tender for fear of not being asked again. Therefore, they enter a 'cover price', which is a tender figure sufficiently high as to virtually guarantee that they will not be considered for this project. This is deemed to be a form of bid rigging.

Bid suppression occurs when a particular company agrees not to submit a bid so another can win the contract. So, company A, instead of submitting an artificially high bid (cover price), agrees with company B not to submit a bid at all, thus increasing the chances of company A to win the bid. This suppression of company A's bid may be in exchange for some sort of arrangement where company B sub-contracts portions of the job back to company A after winning the bid, or the arrangement is reversed on another project. By keeping the low bidder out of the process, the winning tender will be artificially inflated.

Bid rotation and *market allocation* are both collusive agreements between members of a cartel regarding who is going to win which contract. This clearly needs each firm to communicate their own bid to each other before submission to the project. This is the most obvious example of collusion and anti-competitive practice.

Ethical behaviour means that customers, suppliers, competitors and fellow employees should always be treated with respect and fairness. The above examples are clearly unfair business practices in that they attempt to manipulate, misrepresent and abuse commercial information. Not only is this unethical behaviour, in many jurisdictions it would also be illegal within the criminal law, so the risks are high.

16.2.2 Anti-bribery and corruption (ABC)

Bribery is the offering, giving, receiving or soliciting of something of value (usually money or hospitality) for the purpose of influencing the action of others in the discharge of their duties. It is the expectation of a particular return that makes the difference between a bribe and merely a private demonstration of goodwill. It is quite common to offer diaries, calendars and other trivial items to clients – which is accepted as part of general company marketing, goodwill or business courtesies, and items such as these are unlikely to persuade them in their business decisions.

However, to offer or provide excessive or lavish gifts in order to affect their business decisions is known as seeking *undue Influence* over that person's actions and will generally create a perception of impropriety.

The UK government's guidance on the Bribery Act 2010 is quite clear regarding 'facilitation payments' and states:

> Facilitation payments, which are payments to induce officials to perform routine functions they are otherwise obligated to perform, are bribes. There was no exemption for such payments under the previous law nor is there under the Bribery Act.

Regardless of who initiates the deal – the party giving or the party seeking – either party to an act of bribery can be found guilty of the crime independently of the other. Wisely, many companies maintain a record of gifts and hospitality given by them to potential customers and also maintain a list of examples of prohibited gifts. This allows a good deal of protection to appropriate staff.

16.2.3 Negligence

As mentioned in Chapter 12, section 12.1.3, negligence is part of the law of tort where the expected level of care and conduct has not been provided, which results in loss to the client. Professional negligence is a subset of the general rules on negligence to cover situations in which the service provider has represented themselves as having a higher level or specialist set of skills and abilities. The normal rules of negligence rely on establishing that a duty of care is owed by the service provider to the client and there has been a breach of that duty. The standard test of a breach is whether the service provider has matched the abilities of a normal practitioner in their field of expertise. Therefore, negligence in professional services has a higher 'bar' than negligence in other fields, because of the expectation of specialised knowledge.

In terms of professional ethics, practitioners should be aware of this higher level of expectation and conduct themselves accordingly.

16.2.4 Fraud

Fraud is the deliberate deception of a victim in order to secure unfair or unlawful gain. Fraud is also relatively unique in that it is covered in both the civil law (i.e. a fraud victim may sue the perpetrator for damages and compensation) and also the criminal law (i.e. liable to be prosecuted by the appropriate authorities and possibly subject to imprisonment). Fraudulent behaviour is normally carried out for monetary gain, but may also be for other non-pecuniary benefits, such as obtaining a job by issuing false statements, or by claiming qualifications which a candidate does not have, or specious references from former employers. The last two examples are, unfortunately, only too common in the construction industry.

16.2.5 Dishonesty

Dishonesty is an overarching concept which, by definition, means a lack of honesty and also a lack of integrity. Both of these are crucial to professional service providers and also any organisations whose objectives are to provide solutions for clients' requirements. This clearly includes construction companies. As well as being a professional ethical obligation to be honest in all transactions, it is also invariably a legal requirement.

16.2.6 Unfairness

Under US law, the concept of fair dealings is enshrined in anti-trust legislation, which is central to a competitive economy. Employees and staff are therefore expected to deal fairly with customers, suppliers, competitors and other employees. Nobody should take unfair advantage of anyone else through manipulation, concealment, abuse of privileged information, misrepresentation of material facts or any other unfair practice.

16.3 Professional rights and responsibilities

16.3.1 Conflict of interests

A conflict of interest occurs when putting, or giving the appearance of putting, personal, commercial or other interests ahead of the interests of the employer while undertaking employment duties. Personal interests have a habit of clouding judgement and therefore making it difficult to come to good business decisions. Conflicts of interest may also arise when an employee, or a family member, receives improper personal benefit as a result of business connections to the company.

Examples of possible conflicts of interest are listed below.

- using company time and / or resources and / or influence to promote personal interests or the interests of third parties;
- holding a second job with another company with whom your employer conducts business;
- conducting business with companies in which you have an interest;
- serving as a director, officer, etc. with your employer's business partners or competitors;
- holding a significant investment in competitors or in companies with whom the company does business;
- accepting tips or gifts from customers, vendors or other third parties.

Employees (and immediate family members) are normally required to disclose or avoid any activity or interest that may be regarded as a possible conflict interest.

16.3.2 Confidentiality

In any commercial transaction, especially where the values may run into millions of dollars, the need for confidentiality is self-evident. Certainly all project commercial information will be classed as private or restricted and, in modern high tech projects, much of the technical data may be similarly classified, especially if it is still under patent. Confidentiality is about the professional acting with discretion in keeping that information secret and out of the public domain.

Most (if not all) professional bodies impose an ethical duty on their members to respect the confidentiality of information that they acquire as a result of their professional and business relationships. This means that they must not disclose any confidential information to third parties without proper and specific authority unless, of course, they are required to do so by the competent legal authorities.

Confidentiality is therefore at the centre of the relationship of trust between any professional and their client. A commercial manager / cost manager has access to considerable amounts of sensitive commercial information about the project and the business as a whole and has a legal and ethical duty to keep this information confidential unless either the client consents to its disclosure, the law requires disclosure or it is in the public interest. On this last point, many companies also operate policies on whistleblowing – there is a very careful balance between choosing to 'blow the whistle' and maintaining confidentiality.

Case study on professional ethics

An experienced chartered engineer has been sent overseas by his employers to be the new country manager of a declining branch office, with the knowledge that if he 'turns the branch office around' i.e. makes an appropriate level of profit, within six months, then he is likely to be rewarded with a significant promotion. However, when it comes to getting things done within the culture of the country, there are ongoing difficulties with permits and other government bureaucracy, including bringing in much-needed expatriate specialists.

The local sponsor then informs the new manager that everything will flow much more smoothly if a 'lubrication' payment of approximately US$5,000 were to be made to a senior official in the local office of the Ministry. Moreover, it has been made clear that if the money is not forthcoming, the delays will grow increasingly longer.

The new manager is not at all happy about this and makes discreet enquiries with the following results. First, the sponsor has reported the situation correctly and the official, who does effectively control such permits, does indeed want the payment. In addition, the official is related through marriage to the Minister for Development and also the Minister of Immigration. Offering such inducements is also viewed rather quaintly as a local custom. However, whilst it is illegal to offer and accept such monies, the sponsor assures the manager that no-one has been prosecuted in recent memory. However, there is another complication in that if any such payments were made public in the UK, the company would suffer considerable embarrassment, which would obviously affect his chances of the promotion, if not his entire career.

The new manager has found out that his presentation for a multi-million dollar commission was received very favourably, and if this contract was awarded, would be seen as a major coup for both his company and other associated companies in the region. The Commercial section of the Embassy has been working very hard to bring work like this to its national firms. The knock-on effect for other companies would therefore be enormous.

The manager is in charge of the entire operation and has authority to sign off this level of expenditure although ultimately responsible to the Directors in the main headquarters.

After careful thought and some considerable anguish the new manager decides to pay the bribe.

Case study on whistleblowing

In October 2012, 51-year-old Michael Woodford was dismissed as Chief Executive from the Olympus board after merely two weeks into the job, after he questioned payments made by the company. After challenging directors, Woodford called a meeting of the whole Olympus board to discuss the crisis, at which they voted to dismiss him. Within hours he had given all the details to the press and the police but the Olympus board vigorously and publicly denied Woodford's allegations. The board's position was supported by Japanese shareholders who stayed loyal despite the growing amount of evidence that was starting to appear in the public domain about the fraudulent payments. Over the following weeks, regulators uncovered a £1 billion corporate accounting fraud at the Japanese electronics giant, stretching back over more than a decade. The scandal eventually led to the resignation of the entire board and the arrests of several of its most senior former executives, including its chairman. The personal cost to Woodford was enormous. Over a terrifying few months he faced bankruptcy, the collapse of his marriage and a mental breakdown, and at times has said he feared for his own life and that of his family. Woodford received an out-of-court settlement from Olympus of £10 million for his claim for unfair dismissal and discrimination.

What would you have done?

LETTER OF UNDERTAKING
pertaining to
PRINCIPLES OF BUSINESS CONDUCT

To Employer,
Address 1
Address 2

Date

Dear Sir,

In consideration for the release of Tender Documents to us in respect to Tender No . . . ____, I hereby represent acting in my capacity as ________ (Title) of________ (Name of Company), a company organised and existing under the laws of______ (Country of Incorporation) ("the Company"), that

i the Company and its directors and employees understand and accept that the Employer is committed to conducting its business in accordance with the highest standards of ethics and integrity, and that

ii the Company and its officers and employees will carry out business in a lawful, transparent and commercially fair manner.

The Company represents and warrants that it operates and implements policies and procedures to ensure that the Company, its employees, contractors, consultants and other agents and representatives act with the utmost professionalism, honesty and transparency in accordance with international best practice and business ethics in all of its activities and that it has systems and procedures in place to monitor compliance with such policies.

The Company further represents and warrants that prior, during or after the award of any contract, the Company and its directors, employees and other authorised representatives, will:

- Treat all personnel with fairness, dignity, and respect, free from harassment and discrimination; and
- Act in accordance with the highest standard of ethics and best practices prevailing in the relevant profession or industry; and
- Not improperly influence any decision or action or inaction; and
- Not improperly gain access to the Employer's confidential information; and
- Not improperly further the Company's or Company personnel's interests or the interests of the Company's parent company, shareholders, affiliates, sister companies, joint ventures or any other company or entity in which the Company or Company personnel have any interest, to the perceived or actual detriment of and its affiliates or inconsistent with the purpose of the contract; and
- Not make or offer, with respect to the matters which are the subject of the tender, any compensation, commission, agency fee, introduction fee, payment, gift, promise or advantage to any company or person where such payment or advantage would violate applicable laws and regulations; and
- Not give, or receive from, any employee anything of more than nominal value; and
- Not offer, promise, give, pay, request or accept any bribe, gift or inducement in the form of money or otherwise to / from any of Employer's officers, directors, employees and any other authorised representatives, for the purpose of inducing him / her to act contrary to the known rules of honesty and integrity, or influencing his / her behaviour for improperly obtaining or receiving favourable treatment; and
- Immediately notify the Employer when they become aware of any violation of the above.

The Company understands and agrees that any violation of any of the principles mentioned above will be treated as a material breach of the duties assumed under this letter of undertaking, and acknowledges that should the Company be found in violation of the said principles, the Employer may take any possible action available to it under any contract or applicable laws and regulations. This action may include, without limitation, suspending and / or disqualifying the Company in the relevant Tender and prohibiting the Company from participating in further Tenders with the Employer.

The Company acknowledges and agrees that the Employer, acting directly or through its authorised representative(s) or agent(s), shall have the right to conduct an audit with respect to the matters which are the subject of this letter of undertaking. Should the Employer elect to exercise this right, the Company shall, at its own cost, do all things and give all assistance and co-operation to the Employer as may be reasonably requested by them in order to conduct the said audit. This includes, without limitation, (i) making available all books, records, accounts, correspondence, instructions, receipts and memoranda of the Company insofar as they are pertinent to the subject of this letter of undertaking (and the Employer shall have the right to photocopy or otherwise reproduce, at its own cost, any of these documents), and (ii) making the Company's officers and employees, and existing or prospective sub-consultants, available to meet with the Employer or their authorised representative(s) or agent(s) to provide information and / or explanations as may be requested. The Company shall also make all reasonable efforts to obtain a similar undertaking from its existing or prospective sub-consultants, for its and the Employer's benefit. Notwithstanding the Employer's right to see all information described above, it will not carry out an audit of the make-up of the Company's Tender.

(continued)

(continued)

If the Company refuses to comply with this request for an audit, the Employer, in its sole discretion, may take appropriate action against the Company, which may include, without limitation, suspending and/or disqualifying the Company in the relevant Tender and prohibiting the Company from participating in further Tenders with the Employer.

Should the Company breach this letter of undertaking, the Employer may disclose any information provided by the Company, whether provided in confidence or otherwise, to any of the companies in which the Employer has a participating interest.

The Company hereby undertakes, warrants and represents that it will indemnify the Employer against any and all damages, losses, costs, expenses, charges and other disbursements associated with any breach of this letter of undertaking, including, for the avoidance of doubt, consequential losses.

This letter shall be governed and construed in accordance with the laws of ___________ and all disputes arising out of or in connection with this letter shall be subject to the exclusive jurisdiction of the courts of __________.

Signed	:	....	Date	:	....
Name	:	....			
Position	:	....	Company	:	....
Address	:	....	(Company Seal)	:	....

16.4 Summary and tutorial questions

16.4.1 Summary

Professional ethics and codes of conduct are very important in roles that require a fiduciary relationship with a client. The PM / engineer is often acting as the employer's agent or employer's representative and therefore take decisions as though they were the employer. This clearly imposes a very strict requirement that the employer's interest must come before the interest of the consultant's company.

The sections in this chapter are taken from the rules and guidelines given by professional institutions such as the RICS, RIBA, PMP etc. and outline the various issues which need to be considered when representing a client. Many of the issues are also covered by the civil (and criminal) law in many countries, but the fundamental issue is about behaviour as well as actions.

Codes of conduct also give guidance regarding where to draw the line of acceptability. For example, is it all right to accept gifts up to $100? Or nothing at all? Would a gift of $100 affect your judgement as to who is awarded a contract? Some may say yes and some may say no – therefore a written code is essential to avoid any doubt, ambiguity or confusion.

This is also an area where case studies are invaluable – what would you have done in the same circumstances?

16.4.2 Tutorial questions

1 Why are professional services particularly sensitive to ethical issues?
2 Outline the major forms of collusion in the construction industry and what can be done to prevent them in practice.

3 Why should professional negligence be included in a discussion on ethics?
4 Outline five specific circumstances which could be described as conflicts of interests in the construction industry.
5 Read the case study on professional ethics. What would you have done?
6 Read the case study on whistleblowing. What would you have done?
7 Read the letter of undertaking pertaining to principles of business conduct. Is this enforceable on a practical level?

References and further reading

Bew, M. and Richards, M. (2008) Bew-Richards BIM Maturity Model.
British Standards Institute. BS 6079-1:2010 *Project Management. Principles and Guidelines for the Management of Projects*, HMSO.
Centre for Education in the Built Environment (CEBE) (2004) *Effective Management of Contract Variations*, CEBE.
Chartered Institute of Building (CIOB) (2014) *Code of Practice for Project Management*, fifth edition, CIOB.
Greenhalgh, B. D. (2013) *Introduction to Estimating for Construction*, Routledge.
Greenhalgh, B. D. and Squires, G. (2011) *Introduction to Building Procurement*, Routledge.
HMSO (1994) *Constructing the Team. The Final Report of the Government / Industry Review of Procurement and Contractual Arrangements in the UK Construction Industry* (The Latham Report), HMSO.
HMSO (1996) Arbitration Act 1996, HMSO.
HMSO (1996) (2011) Housing Grants, Construction and Regeneration Act 1996 (with amendments 2011), HMSO.
HMSO (1999) Contracts (Rights of Third Parties) Act 1999, HMSO.
HMSO (2015) The Construction (Design and Management) Regulations 2015, HMSO.
Hughes, K. (2013) *Understanding the NEC3 ECC Contract*, Routledge.
Institution of Civil Engineers (2012) *CESMM4: The Civil Engineering Standard Method of Measurement*, fourth edition, Thomas Telford.
Royal Institution of Chartered Surveyors (RICS) (1988) *SMM7: Standard Method of Measurement of Building Works*, seventh edition, RICS.
Royal Institution of Chartered Surveyors (RICS) (2010 and 2013) *New Rules of Measurement* (NRM), RICS.
Royal Institution of Chartered Surveyors (RICS) (2015) *Regulations on Ethics and Professional Standards*, RICS.
Society of Construction Law (SCL) (2002) *Delay and Disruption Protocol*, SCL.
Society of Construction Law (SCL) (2015) *Rider 1 to Delay and Disruption Protocol*, SCL.
Totterdill, B. (2006) FIDIC Users' Guide: A Practical Guide to the 1999 Red and Yellow Books, Thomas Telford.
Tuckman, Bruce (1965) 'Developmental sequence in small groups'. *Psychological Bulletin* 63 (6): 384–99.
UNCITRAL (1958) Convention on the Recognition and Enforcement of Foreign Arbitral Awards (the New York Convention), UNCITRAL.
UNCITRAL (1985) UNCITRAL Model Law on International Commercial Arbitration, with amendments as adopted in 2006.

Useful websites

Chartered Institute of Building – www.ciob.org.uk
Society of Construction Law – www.scl.org.uk
Royal Institution of Chartered Surveyors – www.rics.org

Index

'f' denotes figure and 't' denotes table